ABRASION RESISTANCE OF MATERIALS

Edited by **Marcin Adamiak**

Abrasion Resistance of Materials
Edited by Marcin Adamiak

Published by InTech
Janeza Trdine 9, 51000 Rijeka, Croatia

Publishing Process Manager Masa Vidovic

Technical Editor Teodora Smiljanic

Cover Designer InTech Design Team

Additional hard copies can be obtained from orders@intechopen.com

Abrasion Resistance of Materials, Edited by Marcin Adamiak
p. cm.
ISBN 978-953-51-0300-4

For EU product safety concerns:
IN TECH d.o.o., Prolaz Marije Krucifikse Kozulić 3, 51000 Rijeka, Croatia,
info@intechopen.com or visit our website at intechopen.com.

Contents

Chapter 1 **Abrasion Resistance of Polymer Nanocomposites – A Review 1**
Giulio Malucelli and Francesco Marino

Chapter 2 **Abrasive Effects Observed in Concrete Hydraulic Surfaces of Dams and Application of Repair Materials 19**
José Carlos Alves Galvão, Kleber Franke Portella and Aline Christiane Morales Kormann

Chapter 3 **Abrasion Resistance of High Performance Fabrics 35**
Maja Somogyi Škoc and Emira Pezelj

Chapter 4 **Numerical Simulation of Abrasion of Particles 53**
Manoj Khanal and Rob Morrison

Chapter 5 **Low Impact Velocity Wastage in FBCs – Experimental Results and Comparison Between Abrasion and Erosion Theories 75**
J. G. Chacon-Nava, F. Almeraya-Calderon, A. Martinez-Villafañe and M. M. Stack

Chapter 6 **Heat and Thermochemical Treatment of Structural and Tool Steels 99**
Jan Suchánek

Chapter 7 **Analysis of Abrasion Characteristics in Textiles 119**
Nilgün Özdil, Gonca Özçelik Kayseri and Gamze Süpüren Mengüç

Chapter 8 **Rubber Abrasion Resistance 147**
Wanvimon Arayapranee

Chapter 9 **Effect of Abrasive Size on Wear 167**
J. J. Coronado

Chapter 10 **Abrasion Resistance of Cement-Based Composites 185**
Wei-Ting Lin and An Cheng

1

Abrasion Resistance of Polymer Nanocomposites – A Review

Giulio Malucelli and Francesco Marino
Politecnico di Torino, DISMIC
Italy

1. Introduction

In order to be suitable for tribological applications, polymeric materials, which can usually exhibit mechanical strength, lightness, ease of processing, versatility and low cost, together with acceptable thermal and environmental resistances, have to show good abrasion and wear resistance. This target is not easy to achieve, since the viscoelasticity of polymeric materials makes the analysis of the tribological features and the processes involved in such phenomena quite complicated.

Indeed, it is well-known that an improvement of the mechanical properties can be effectively achieved by including "small" inorganic particles in the polymer matrices (Dasari et al., 2009).

For applications taking place in hard working conditions, such as slide bearings, the development of composite materials, which possess a high stiffness, toughness and wear resistance, becomes crucial. On the one hand, the extent of the reinforcing effect depends on the properties of composite components, and on the other hand it is strongly affected by the microstructure represented by the filler size, shape, homogeneity of distribution/dispersion of the particles within the polymer, and filler/matrix interface extension. This latter plays a critical role, since the composite material derives from a combination of properties, which cannot be achieved by either the components alone.

Thus it is generally expected that the characteristics of a polymer, added of a certain volume fraction of particles having a certain specific surface area, are more strongly influenced when very small particles (*nanofillers*), promoting an increased interface within the bulk polymer, are used (Bahadur, 2000; Chen et al., 2003; Karger-Kocsis & Zhang, 2005; Li et al., 2001; Sawyer et al., 2003). However, this happens only when a high dispersion efficiency of the nanoparticles within the polymer matrix is assessed: indeed, nanoparticles usually tend to agglomerate because of their high specific surface area, due the adhesive interactions derived from the surface energy of the material. In particular, the smaller the size of the nanoparticles, the more difficult the breaking down of such agglomerates appears, so that their homogeneous distribution within the polymer matrix is compromised.

As a consequence, the development of nanocomposites showing high tribological features requires a deep investigation on their micro-to-nanostructure, aiming to find synergistic mechanisms and reinforcement effects exerted by the nanofillers (Burris et al., 2007).

In addition, the way in which nanofillers can improve the tribological properties of polymers depends on the requirement profile of the particular application, i.e. the friction coefficient and the wear resistance cannot be considered as real material properties, since they depend on the systems in which these materials have to function.

In particular, such applications as brake pads or clutches usually require a high friction coefficient and, at the same time, a low wear resistance; however, in other circumstances (like in the case of gears or bearings, acting as smooth metallic counterparts under dry sliding conditions) the development of polymer composites having low friction and wear properties is strongly needed.

The abrasion performances of polymeric materials depend on several factors, such as the wear mechanisms involved, the abrasive test method used, the bulk and surface properties of the tested specimens, ...

Many papers reported in the literature focus on the investigation on the physical processes involved in abrasive wear of a wide variety of polymers; the obtained results demonstrate that two very different mechanisms of wear may occur in polymers, namely cohesive and interfacial wear processes, as schematically shown in Figure 1.

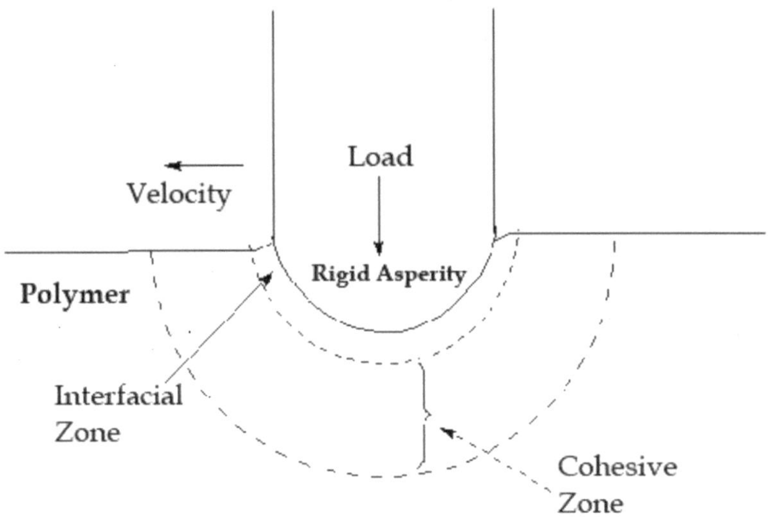

Fig. 1. Schematic representation of cohesive and interfacial wear processes (Adapted from Briscoe & Sinha, 2002)

In the cohesive wear processes, such as abrasion wear, fatigue wear and fretting, which mainly depend on the mechanical properties of the interacting materials, the frictional work involves quite large volumes close to the interface, either exploiting the interaction of surface forces and the consequent traction stresses or through the geometric interlocking exerted by the interpenetrating contacts. Contact stresses and contact geometry represent two key parameters that determine the extent of such surface zone.

On the other hand, the frictional work in interfacial wear processes (like transfer wear, chemical or corrosive wear) is dissipated in much thinner zones and at greater energy density with respect to cohesive wear processes, so that a significant increase in local temperature occurs. Furthermore, the extent of wear damage can be substantially ascribed to the chemistry of the surfaces involved, rather than to the mechanical properties of the interacting materials.

As far as cohesive processes are concerned, *abrasion wear*, which is the most common type of wear encountered in polymer composites, can be divided into two-body and three-body abrasion wear. The former occurs in the presence of hard asperities that plough and induce plastic deformation or fracture of the softer asperities.

The latter relates to the presence of hard abrasive particles or wear debris in between the sliding bodies: such particles or debris derive from environmental contaminants or can be the consequence of two-body abrasion processes. In general abrasion wear depends on several factors, such as the hardness of the materials in contact, the applied load and sliding distance and the geometry of the abrasive particles as well.

Fatigue wear derives from surface fatigue phenomena, i.e. from the repeated stressing and un-stressing of the contacts, and can lead to fracture through the accumulation of irreversible changes, which determine the generation, growth and propagation of cracks. This kind of wear may also occur together with *delamination wear*, where shear deformations of the softer surface, caused by traction of the harder asperities, promote the nucleation and coalescence of subsurface cracks. As a consequence, the delamination (i.e. detachment) of fragments having larger size occurs.

Fretting wear is caused by relative oscillatory motions of small amplitude taking place between two surfaces in contact. The produced wear fragments can either escape from between the surfaces, thus promoting a fit loss between the surfaces and a decrease of clamping pressure, which may lead to higher vibration effects, or remain within the sliding surfaces, so that pressure increases and seizure eventually occurs.

Transfer wear belongs to interfacial wear processes and involves the formation of a transfer film (solid or liquid, depending on the interfacial temperature) in polymer-metal, polymer-ceramic, polymer-polymer sliding contacts. Such film invariably transfers from polymer to metal or ceramic, whereas the direction of transfer is not obvious in the case of polymer-polymer sliding contacts.

Several parameters can influence the formation of the transfer film and its role on the subsequent wear processes: thickness and stability of the film, cohesion features between the transfer layers, adhesion forces between the film and the sliding counterpart, chemical reactivity and surface roughness of the counterface slider, polymer structure (crystallinity, flexibility, presence of pendant groups or side chains, …), adopted sliding conditions (temperature, normal load, velocity, atmosphere, …) and presence of fillers.

Chemical wear involves a chemical reaction (oxidation, degradation, hydrolysis, …, which lead to polymer chain scission with the subsequent MW decrease) in between the sliding bodies or a material in itself or a material with the surrounding environment.

A schematic representation of the basic tribological interactions leading to wear particle generation is depicted in Figure 2.

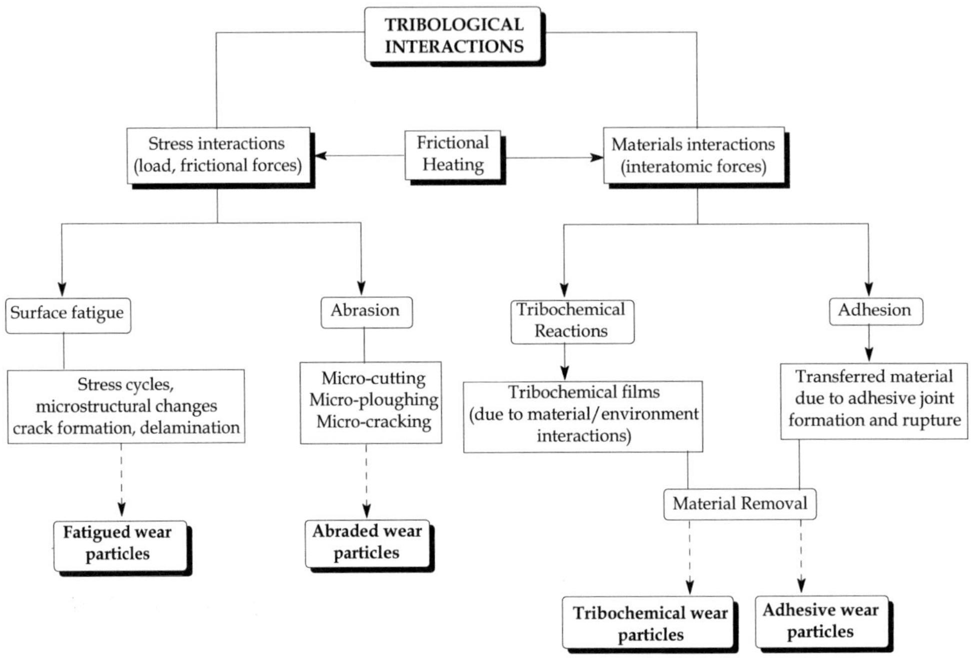

Fig. 2. Different wear processes leading to the formation of material particles (adapted from Czichos, 2001)

It is worthy to note that the wear mechanisms in polymer systems described above for macro- and micro-levels are quite different from those encountered at nano-level.

First of all, nano-level involves very low applied loads (from μN to nN); in addition, the wear particle generation is negligible and the original surface topography is more likely to be preserved for an extended period because of the adopted low wear rate.

Other differences concern the friction forces involved at the nano-level, since the ploughing factor and the inertial effect of the moving components are different, as well as the role exerted by surface forces (adhesion and electrostatic interactions), which become very important.

In the following paragraphs, a review on the recent studies on the tribological behavior of thermoplastic nanocomposites is presented. The role of the structure of the nanofillers and of their morphology (aspect ratio, effectiveness of dispersion within the polymer, …) and the possible interactions with the environment are widely discussed.

2. Tribology of thermoplastic nanocomposites

2.1 PEEK-based nanocomposites

Poly(ether ether ketone) (PEEK) is a high performance injection mouldable thermoplastic that can be widely used for many applications that require high mechanical strength and an outstanding thermo-mechanical stability.

This polymer has a high glass transition temperature (Tg≈143°C) and a high melting point (Tg≈343°C) and it is also regarded as one of the most promising polymer materials for tribological applications in aqueous environments.

Nevertheless, it seems that neat PEEK exhibits relatively poor wear resistance with water lubrication in some cases, so that different types of fillers (and nanofillers) have been incorporated into this polymer, aiming to facilitate more applications by enhancing its anti-wear features. In particular, short carbon fibers (SCFs) are currently used in PEEK-based composites for improving its wear resistance, even at elevated temperatures and under aqueous conditions (water lubrication).

Very recently, Zhong investigated the tribological properties of PEEK/SCF/zirconia composites under aqueous conditions, using a three-pin-on-disc configuration (Zhong et al., 2011). A synergistic effect of SCFs with zirconia nanoparticles was assessed: indeed, the composites showed excellent wear resistance under aqueous conditions; SCFs were found to carry the main load between the contact surfaces and to protect the polymer matrix from further severe abrasion of the counterpart. Nano-ZrO_2 efficiently inhibited SCF failure either by reducing the stress concentration on the CF interface through reinforcement of the matrix or by lowering the shear stress between the sliding surfaces via a positive rolling effect of the nanoparticles between the material pairs.

Werner et al. investigated the influence of vapour-grown carbon nanofibres (CNFs) on the wear behaviour of PEEK (Werner et al., 2004). To this aim, unidirectional sliding tests against two different counterpart materials (100Cr6 martensitic bearing steel and X5CrNi18-10 austenitic stainless steel) were performed on injection moulded PEEK-CNF nanocomposites. CNFs were found to reduce the wear rate of PEEK very significantly, as compared to a variety of commercial PEEK grades. This behaviour was attributed to CNFs, which act as solid lubricants; in addition, the roughening effect on the counterpart exerted by CNFs, because of their small size, was minimised with respect to conventional fibre fillers (carbon fibres, PAN-based carbon fibres, glass fibres).

McCook and coworkers investigated the role of different micro and nanofillers on the tribological properties of PEEK in dry sliding tests against 440C stainless steel counterfaces (McCook et al., 2007). To this aim, microcrystalline graphite, carbon nano-onions, single-walled carbon nanotubes, C60 carbon fullerenes, microcrystalline WS_2, WS_2 fullerenes, alumina nanoparticles and PTFE nanoparticles were jet-milled with PEEK and the friction coefficients and wear rates of the obtained composites were measured in open laboratory air (45% R.H.) and in a dry nitrogen environment (less than 0.5% R.H.).

Both wear rate and friction coefficient were reduced in the dry nitrogen environment: in particular, the more wear resistant coatings also had lower friction coefficients. On the contrary, in open air environments the more wear resistant coating exhibited the higher friction coefficients. Furthermore, the polymeric nanocomposites investigated showed similar environmental responses, regardless of the type of micro or nanofillers used.

Hou and coworkers performed tribological ball-on-flat sliding wear tests on PEEK-based nanocomposites incorporating inorganic fullerene-like tungsten disulfide nanoparticles (Hou et al., 2008). The friction coefficient was found to decrease about 3 times in the presence of 2.5 wt.% nanoparticles, with respect to the neat PEEK: this behaviour was attributed to the lubricating capability of the nanofillers.

Zhang et al. investigated the effect of nano-silica particles on the tribological behaviour of PEEK: silica nanoparticles were compounded with the polymer by means of a ball milling technique (Zhang et al., 2008). The wear resistance of PEEK was significantly improved after incorporating nano-SiO_2 and at a rather low filler loading (1 vol.%), the composites showed the optimum wear resistance, which was ascribed to the reduced perpendicular deformation of PEEK matrix and to the decreased tangential plastic flow of the surface layer involved in friction processes. Furthermore, the nanocomposites evidenced much smoother surfaces with respect to neat PEEK.

Pursuing this research, the role of the same nano-silica particles on the tribological behaviour of SCF-reinforced PEEK was also investigated (Zhang et al., 2009). To this aim, 1 vol.% (1.51 wt.%) nano-SiO_2 particles were compounded with SCF/PTFE/graphite filled PEEK in a Brabender mixer; the obtained composite materials were tested using a block-on-ring apparatus at room temperature (counterpart: 100Cr6 steel ring), in extremely wide pressure and sliding velocity ranges. Under all the conditions investigated, nano-SiO_2 particles remarkably reduced the friction coefficients; above 2 MPa pressures, the nanoparticles were found to reduce the wear rate: this behaviour was attributed to a protection effect of SCF/PEEK interface exerted by the nanoparticles, which are able to reduce the stress concentration on SCFs taking place in the surface layer involved into friction.

Zhang also investigated the effect of different amounts of nano-silica particles on the tribological behavior of SCF-reinforced PEEK composites. The nanoparticle loading was varied from 1 to 4 vol.% (Zhang et al., 2009).

The variation of nanoparticle content from 1 to 4 vol.% did not significantly affect the friction coefficients of the nanocomposites; in addition, operating with low pressure-sliding velocity (pv) factors, the nanoparticles turned out to worsen the wear rate of the composites, because of the abrasion on SCFs exerted by nanoparticle agglomerates. On the contrary, with a high pv factor, such agglomerates were crushed into tiny ones, so that nano-silica particles were capable to protect SCFs reducing their failures. Similar wear rates were found for the nanocomposites tested at very high pv factors.

2.2 Polyolefin-based nanocomposites

Thermoplastic polyolefins like poly(ethylene)s (PEs) and poly(propylene) (PP) are well-established polymers available at the market, each having a different structure and very different behaviour, performances and applications (Feldman & Barbalata, 1996). Several papers deal with their tribological properties, in the presence of different types of nanofillers.

High density poly(ethylene) (HDPE) was used as matrix for preparing nanosilica coatings, the wear resistance of which was measured using a rotative drum abrader (Barus et al., 2009). It was found that this parameter, despite a significant increase in the mechanical properties of the nanocomposites (stiffness, yield strength and fracture toughness), exhibited lower values with respect to the neat polymer.

Johnson and coworkers manufactured and tested the wear behaviour of HDPE/multi-walled-carbon-nanotubes composites (Johnson et al., 2009). Different weight percentages of nanotubes (1, 3 and 5%) were used for preparing the samples, which were tested on a block-on-ring apparatus. Wear resistance and frictional properties of HDPE were found to improve in the presence of the nanofillers; furthermore, the addition of multi-walled-carbon-nanotubes to HDPE turned out to bring wear rates down to the level seen in ultra-high molecular weight poly(ethylene) (UHMWPE).

The effect of the presence of Alumina nanoparticles (5 wt.%) was exploited for investigating the abrasion resistance of low-density poly(ethylene) (LDPE)-based nanocomposites (Malucelli et al., 2010). The abrasion resistance of the nanocomposites increased in the presence of the nanofillers, as indicated by the decrease of the Taber Wear Index with respect to the neat polymer.

Very recently, Xiong and coworkers investigated the effect of the presence of nano-hydroxyapatite (nano-HAP) on the tribological properties of non-irradiated and irradiated UHMWPE composites, prepared by using a vacuum hot-pressing method (Xiong et al., 2011). The friction coefficients and wear rates were measured by using a reciprocating tribometer (counterface: CoCr alloy plates). The presence of 7 wt.% nano-HAP in the polymer matrix resulted in lowering both the friction coefficients and wear rate, irrespective of using irradiated or non-irradiated samples, whereas filling 1 wt.% nano-HAP reduced friction coefficients and wear rate of the non-irradiated UHMWPE only.

Misra and coworkers investigated the tribological behaviour of polyhedral oligomeric silsesquioxanes (POSS)/poly(propylene) nanocomposites (Misra et al., 2007). The relative friction coefficient of the samples turned out to strongly decrease from 0.17 for neat PP to 0.07 for the nanocomposite containing 10 wt.% POSS: this behaviour was ascribed to the increase of the surface hardness and of the modulus, due to the presence of the nanofiller.

2.3 Fluorinated-based nanocomposites

Fluorinated polymers usually exhibit many desirable tribological features, including low friction, high melting temperature and chemical inertness. However, their anti-wear applications have been somewhat limited by their poor wear resistance, which has led to the failure of anti-wear components and films.

Therefore, many researchers have tried to reinforce fluorinated polymers using different fillers, such as glass fibres, carbon fibres, ceramic powders, non-ferrous metallic powders: unfortunately, these fillers induced a large frictional coefficient and abrasion. Quite recently, nanometer size inorganic powders have been chosen as fillers capable to enhance the wear behaviour of fluorinated polymers.

Poly(tetrafluoroethylene), PTFE, is the most common fluorinated polymer used for tribological purposes.

Lee and coworkers added carbon-based nanoparticles, synthesized by heat treatment of nanodiamonds, to PTFE, in order to prepare fluorinated nanocomposites (Lee et al., 2007). The wear resistance, measured through ball-on-plate wear tests, was found to depend on the heat treatment, which nanodiamonds were subjected to: in particular, wear resistance turned out to increase when nanodiamonds were heated at 1000°C. Beyond this temperature, carbon nanoparticles became aggregated and therefore the wear coefficient of the obtained nanocomposites increased: this failure in the wear behaviour was ascribed to the formation of carbon onions that promoted the aggregation of carbon nanoparticles.

Single-walled carbon nanotubes have been exploited for lowering the wear rates of PTFE (Vail et al., 2009). A linear reciprocating tribometer was exploited for performing the tests (counterface: 304 stainless steel) on nanocomposite samples containing up to 15 wt.% nanotubes. The obtained results clearly indicated that, in the presence of low nanofiller loadings (5 wt.%), PTFE wear resistance is improved by more than 2000% and friction coefficient increased by ≈50%.

Shi and coworkers have studied the effect of various filler loadings (from 0.1 to 3 wt.%) on the tribological properties of carbon-nanofiber (CNF)-filled PTFE composites (Shi et al., 2007). The friction and wear tests were conducted on a ring-on-ring friction and wear tester. The counterface materials was steel 45.

The obtained results showed that the friction coefficients of the PTFE composites decreased initially up to a 0.5 wt.% filler concentration (during sliding, the released CNFs transfer from the composite to the interface between the mating surfaces, acting as spacers and thus preventing direct contact between the two surfaces and lowering the friction coefficient) and then increased, whereas the anti-wear properties of the materials increased by 1-2 orders of magnitude in comparison with those of PTFE. Finally, the composite having 2 wt.% of CNFs exhibited the best anti-wear properties under all the experimental friction conditions.

The tribological investigation on fluorinated polymers has been also extended to PTFE-based blends, as described by Wang and coworkers (Wang et al., 2006). In particular, Xylan 1810/D1864, a commercially available PTFE blend for dry lubricant and corrosion resistant coatings, has been blended with alumina nanoparticles at different loadings (from 5 to 20 wt.%). The wear resistance was measured using a Taber Abrasion Tester and was found to decrease with increasing the content of the embedded alumina nanoparticles in the polymer matrix. The minimum wear rate was achieved when the nanoparticle loading was 20 wt.%.

Another paper from Burris and Sawyer reports on the role of irregular shaped alumina nanoparticles on the wear resistance of Al_2O_3/PTFE nanocomposites (Burris & Sawyer, 2006). A reciprocating pin-on-disc tribometer was used for testing the wear and friction of the samples (counterface: AISI 304 stainless steel plates). It was found that the inclusion of irregular shaped alumina particles is more effective in reducing PTFE wear than spherical shaped particles (the wear resistance of PTFE was increased 3000x in the presence of 1 wt.% former nanofiller), but also determines an increased friction coefficient.

Another fluorinated polymer, namely poly(vinylidene fluoride), was used as matrix for preparing nanocomposites containing a phyllosilicate (organoclay) by Peng and coworkers (Peng et al., 2009). The friction and wear tests were conducted on different loaded nanocomposites (clay content: 1 - 5 wt.%), using a block-on-ring wear tester (mated ring specimen: carbon steel 45, GB 699-88). The nanoclay at 1-2 wt.% turned out to be effective for improving the tribological properties of neat PVDF, since such filler may act as a reinforcement to bore load and thus decrease the plastic deformation.

Tribological studies were also performed on PTFE-based fabric composites (Sun et al., 2008; Zhang et al., 2009). In particular, Sun and coworkers prepared polyester fabric composites, in order to study the influence of alumina nanoparticles and PTFE micro-powders embedded in an epoxy matrix on the tribological properties of the fabric composites. The excellent tribological performance of the fillers significantly turned out to enhance the wear resistance of the fabric polyester composites.

Zhang et coworkers were able to improve the wear resistance of PTFE/phenolic/cotton fabric composites, by dispersing functionalized multi-walled carbon nanotubes in the phenolic resin (Zhang et al., 2009). Sliding tests were performed on a pin-on-disc tribometer (flat-ended AISI-1045 pin). The high homogeneity of dispersion of the nanofiller allowed to achieve an improved wear resistance in the fabric composites; furthermore, the tribological properties of the obtained systems were found to strongly depend on the carbon nanotubes content: 1 wt.% nanofiller was the optimum loading for maximizing the wear resistance of the fabric composites.

2.4 Poly(amide)-based nanocomposites

Poly(amide) 6 and 66 (Nylon 6 and Nylon 66) have been widely used as engineering plastics in different applications, such as bearings, gears or packaging materials. They possess an outstanding combination of properties such as high toughness, tensile strength and abrasion resistance, low density and friction coefficient and quite easy processing. Indeed, their abrasion resistance is a key factor for their widespread applications.

Aiming to further improve their mechanical properties and tribological behaviour, nylons were reinforced with some micro-particles or fibres, such as CuS, CuF_2, CuO, PbS, CaO, CaS and carbon fibres: they were effective in reducing the wear rate of polyamides (Bahadur et al., 1996).

In quite recent years, as for other thermoplastic matrices, several nano-materials were served as suitable fillers of poly(amides) for improving their integrated properties, particularly referring to their tribological behavior.

Garcia and coworkers found that nano-SiO_2 could reduce effectively the coefficient of friction and wear rate of nylon 6: in particular, the addition of 2 wt.% nano-SiO_2 determined the lowering of the friction coefficient from 0.5 to 0.18 (Garcia et al., 2004). This was possible since the surface of nylon 6 nanocomposites was well protected by the transfer film on the surface of the metal counterface. At the same time, the low silica loading led to a reduction in wear rate by a factor of 140, whereas the effect of higher silica loadings was less pronounced.

Dasari and coworkers reported on the role of nanoclays on the wear characteristics of nylon 6 nanocomposites processed via different routes (Dasari et al., 2005). They demonstrated that aggregated nanoclay particles result in the worst wear resistance of the nanocomposites, whereas the systems, which exhibit a good interfacial adhesion of clay to polymer matrix, together with an homogeneous clay dispersion, determine substantial improvements of wear resistance.

Zhou and coworkers investigated the tribological behaviour of Nylon 6/Montmorillonite clay nanocomposites: the poor abrasion resistance exhibited by the nanocomposites was attributed to the presence of defects at the clay/polymer interface, resulting in lower wear resistance of the polymer matrix as the nanofiller content increased (Zhou et al., 2009).

Sirong and coworkers studied the tribological behaviour of Nylon 66/organoclay nanocomposites, in the presence of styrene-ethylene/butylene-styrene triblock copolymer grafted with maleic anhydride (SEBS-g-MA) as a toughening agent (Sirong et al., 2007). A pin-on-disc friction and wear testing apparatus was used in sliding experiments (counterface: 65 HRC steel disc). It was demonstrated that the use of SEBS-g-MA allows to obtain significant improvements as far as the wear resistance of the nanocomposite is concerned: this behaviour was ascribed to the toughening effect of SEBS-g-MA, which favours the transfer of a uniform, continuous and smooth thin film to the steel counterface, thus avoiding the direct contact of this latter with the nanocomposite.

Poly(amide) 66 was also chosen as matrix for preparing nanoparticle-filled composites (Chang et al., 2006). Different fillers, such as TiO_2 nanoparticles (5 vol.%), short carbon fibres (15 vol.%) and graphite flakes (5 vol.%), were added to the polymer and the obtained composites tested on a pin-on-disc apparatus (counterface: polished steel disc). It was found that nano-TiO_2 could effectively reduce the frictional coefficient and wear rate, especially under higher pv conditions. In order to further understand the wear mechanisms, the worn surfaces were examined by scanning electron microscopy and atomic force microscopy; a positive rolling effect of the nanoparticles between the material pairs was proposed, which contributes to the remarkable improvement of the load carrying capacity of polymer nanocomposites.

Quite recently, Ravi Kumar and coworkers studied the synergistic effect of nanoclay and short carbon fibers on the abrasive wear behavior of nylon 66/poly(propylene) nanocomposites (Ravi Kumar et al., 2009). A modified dry sand rubber wheel abrasion tester was employed for performing the three-body abrasive wear experiments. The obtained results clearly indicated that the addition of nanoclay/short carbon fiber in PA66/PP significantly influences wear under varied abrading distance/loads. Furthermore, it was found that nanoclay filled PA66/PP composites exhibited lower wear rates with respect to short carbon fiber filled PA66/PP composites.

2.5 Poly(oxymethylene)-based nanocomposites

Poly(oxymethylene) (POM) is an engineering polymer that has been widely used as self-lubricating material for many applications, such as automobile, electronic appliance and engineering. This polymer exhibits good fatigue resistance, creep resistance and high impact strength. Its low friction coefficient derives from the flexibility of the linear macromolecular chains; in addition, its high crystallinity and high bond energy result in good wear resistant properties. Some papers report on the preparation of polymeric nanocomposites based on POM.

Various fillers or fibers, such as graphite, MoS_2, Al_2O_3, PTFE, glass and carbon fibers, have been incorporated into POM matrices as internal lubricants or reinforcements to further enhance the tribological properties of such a polymer.

Kurokawa et al. investigated the tribological properties of POM composites containing very small amounts of silicon carbide (SiC) and/or calcium salt of octacosanoic acid (Ca-OCA), as well as PTFE (Kurokawa et al., 2000). It was found that the incorporation of Ca-OCA into POM/SiC composites drastically lowered their friction coefficient; furthermore, the wear rate was also lowered because of the nucleating effect of SiC and Ca-OCA.

Wang and coworkers prepared POM/MoS_2 nanocomposites by in situ intercalation polymerization: the intercalated composites showed a significant decrease of friction coefficient, together with an improved wear resistance, especially under high load, while the heat resistance of the composites decreased slightly (Wang et al., 2008).

The same research group also prepared POM/ZrO_2 nanocomposites, which evidenced better wear resistance with respect to neat POM, whereas the change in friction coefficient of the nanocomposites was very limited. (Wang et al., 2007)

Sun and coworkers studied the tribological properties of POM/Al_2O_3 nanocomposites (Sun et al., 2008). The friction and wear measurements were conducted on a friction and wear tester, using a block-on-ring arrangement (counterface: HRC50-55 plain carbon steel ring). It was found that alumina nanoparticles were more effective in enhancing the tribological properties of Poly(oxymethylene) nanocomposites in oil lubricated condition rather than in dry sliding experiments. Indeed, the former environment allows to form a uniform and compact transfer film on the surface of the counterpart steel ring, whereas the transfer film under dry sliding condition is destroyed by the agglomerated abrasives residing between the friction surfaces. The optimal nanoparticles content in POM nanocomposites was 9% under oil lubricated condition, below which alumina nanoparticles between the friction surfaces were still under saturation.

Sun and coworkers have also investigated the tribological behaviour of Poly(oxymethylene) (POM) composites compounded with nanoparticles, PTFE and MoS_2 in a twin-screw extruder (Sun et al., 2008). The tribological tests were performed on a friction and wear tester using a block-on-ring arrangement under dry sliding and oil lubricated conditions, respectively. The better stiffness and tribological properties exhibited by POM nanocomposites with respect to POM composites were attributed to the high surface energy of the nanoparticles; the only exception was represented by the decreased dry-sliding tribological properties of POM/3%Al_2O_3 nanocomposite, ascribed to Al_2O_3 agglomeration. Furthermore, the friction coefficient and wear volume of POM nanocomposites under oil lubricated condition decreased significantly.

2.6 Poly(methylmethacrylate)-based nanocomposites

Poly(methylmethacrylate), PMMA, is an important engineering polymer, which finds application in many sectors such as aircraft glazing, signs, lighting, architecture, and transportation. In addition, since PMMA is non-toxic, it could be also useful in dentures, medicine dispensers, food handling equipment, throat lamps, and lenses.

Unfortunately, this polymer shows poor abrasion resistance with respect to glass, thus limiting its potential use in other fields. Despite several efforts, attempts to improve the PMMA scratch and abrasion resistance have induced other drawbacks, such as a decrease of the impact strength, so that researchers focused on the preparation of PMMA nanocomposites.

Avella and coworkers studied the tribological features of PMMA-based nanocomposites filled with calcium carbonate ($CaCO_3$) nanoparticles, exploiting in situ polymerization (Avella et al., 2007). In order to improve inorganic nanofillers/polymer compatibility, poly(butylacrylate) chains have been grafted onto $CaCO_3$ nanoparticle surface.

$CaCO_3$ nanoparticles, regardless of the presence of the grafting agent, turned out to significantly improve the abrasion resistance of PMMA also modifying its wear mechanism: indeed, the nanoparticles induced only micro-cutting and/or micro-ploughing phenomena, thus generating a plastic deformation and consequently increasing the abrasion resistance of the polymer matrix.

The same research group also investigated the tribology of PMMA-based nanocomposites containing modified silica nanoparticles, obtained through in situ polymerization approach (Avolio et al., 2010). The high compatibility between silica nanoparticles and the polymer allowed to significantly improve the abrasion resistance of PMMA, because nanoparticles were able to support part of the applied load, thus reducing the penetration of grains of the rough abrasive wheel into PMMA surface and contributing to the wear resistance of the material.

Dong and coworkers prepared Poly(methyl methacrylate)/styrene/multi-walled carbon nanotubes (PMMA/PS/MWNTs) copolymer nanocomposites by means of in situ polymerization method (Dong et al., 2008). The tribological behavior of the copolymer nanocomposites was investigated using a friction and wear tester under dry conditions: with respect to pure PMMA/PS copolymer, the copolymer nanocomposites showed not only better wear resistance but also smaller friction coefficient. MWNTs were found to strongly improve the wear resistance property of the copolymer nanocomposites, because of their self-lubricating features, their homogeneous and uniform distribution within the copolymer matrix and their help in forming thin running MWNTs films that slide against the transfer film (developed on the surface of the stainless steel counterface).

Very recently, Carrion and coworkers exploited single-walled carbon nanotubes modified with an imidazolium ionic liquid for preparing PMMA nanocomposites and studying their dry tribological performances as compared to neat PMMA or to the nanocomposites containing pristine carbon nanotubes without ionic liquid (Carrion et al., 2010). The tribological behavior of the obtained nanocomposites, studied against AISI 316L stainless steel pins, resulted in a significant wear rate decrease with respect to PMMA/carbon nanotubes (-58%) and neat PMMA (-63%).

2.7 Other thermoplastic-based nanocomposites

Some other thermoplastic engineering and specialty polymers have been considered as far as tribological issues are concerned. In the following, we will summarize the recent progress in understanding wear and friction in nanocomposite systems based on these polymers.

Bhimaraj and coworkers studied the friction and wear properties of poly(ethylene) terephthalate (PET) filled with alumina nanoparticles (up to 10 wt.% nanofiller), using a reciprocating tribometer (Bhimaraj et al., 2005). The obtained results showed that the addition of alumina nanoparticles can increase the wear resistance by nearly 2x over the unfilled polymer. Furthermore, the average friction coefficient also decreased in many cases. This behavior was attributed to the formation a more adherent transfer film that protects the sample from the steel counterface, although the presence of an optimum filler content could be ascribed to the development of abrasive agglomerates within the transfer films in the higher wt.% samples.

Another paper from the same research group reports on the effect of particle size, loading and crystallinity on PET/Al_2O_3 nanocomposites (Bhimaraj et al., 2008). The nanocomposite samples were tested in dry sliding against a steel counterface. The tribological properties were found to depend on crystallinity, filler size and loading; in addition, wear rate and friction coefficient were very low at optimal loadings that ranged from 0.1 to 10 wt.%, depending on the crystallinity and particle size.

Wear rate were found to lower monotonically with decreasing particle size and crystallinity at any loading in the range tested.

Poly(etherimide)s (PEIs) are high-performance thermoplastics with high modulus and strength, superior high temperature stability, as well as electrical (insulating) and dielectric properties (very low dielectric constant). These polymers perform successfully in aerospace, electronics, and other applications under extreme conditions. Nevertheless, pure PEIs show such disadvantages as brittleness and high wear rate, which limit their applications. Therefore, appropriate modifications of PEIs with nanofillers have been proposed, in order to widen their industrial applications.

Chang and coworkers reinforced PEI with titania nanoparticles, in the presence of short carbon fibres (SCFs) and graphite flakes as well (Chang et al., 2005). Wear tests were performed on a pin-on-disc apparatus, using composite pins against polished steel counterparts, under dry sliding conditions, different contact pressures and various sliding velocities. SCFs and graphite flakes turned out to remarkably improve both the wear resistance and the load-carrying capacity. Nevertheless, the addition of nano-TiO_2 further reduced the frictional coefficient and the contact temperature of the composites, especially under high pv conditions.

The same research group investigated the role of the presence of nano- or micro-sized inorganic particles (5 vol.% nano TiO_2 or micro-$CaSiO_3$) on the tribological behavior of PEI matrix composites, additionally filled with SCFs and graphite flakes (Xian et al., 2006). The influence of these inorganic particles on the sliding behavior was assessed with a pin-on-disc tester at room temperature and 150°C.

The obtained results showed that both micro and nano particles could reduce the wear rate and the friction coefficient of the PEI composites under the experimental adopted conditions, but in a different temperature range: indeed, the microparticles filled composites showed improved tribological features at room temperature, whereas the nano-titania-filled composites possessed the lowest wear rate and friction coefficient at elevated temperature. The tribological improvements evidenced by the nano-particles were attributed to the formation of transfer layers on both sliding surfaces together with the reinforcing effect.

Very recently, Li and coworkers dispersed carbon nanofibers (from 0.5 to 3 wt.%) in a PEI matrix through a melt mixing method and tested the tribological properties of the obtained nanocomposites (Lee et al., 2010). The composites containing 1 wt.% CNFs showed very high wear rates comparable with that of pure PEI; nevertheless, higher CNF loadings promoted a significant reduction in wear rate at steady state wear.

Like PMMA, also poly(carbonate) (PC), an amorphous engineering thermoplastic, which combines thermal stability, good optical properties, outstanding impact resistance and easy processability, shows poor scratch and abrasion resistance with respect to glass, thus limiting its potential use in fields other than medical, optics, automotive, ...

Carrion and coworkers prepared a new polycarbonate nanocomposite containing 3 wt.% organically modified nanoclay by extrusion and injection moulding, and its tribological properties were measured under a pin-on-disc configuration against stainless steel (Carrion et al., 2008). The obtained nanocomposites showed 88% of reduction in friction coefficient and up to 2 orders of magnitude reduction in wear rate with respect to the neat polymer. Such good tribological performances were attributed to the uniform microstructure achieved and to the nanoclay intercalation.

3. Conclusion

The significant spreading of research activities concerning the tribology of thermoplastics and thermoplastic-based nanocomposites demonstrates that this topic is very up-to-date. Indeed, several low-loading, low-wear polymer nanocomposites are being prepared and evaluated in tribology laboratories.

In many cases, nanocomposite systems result in outperforming traditional macro- and micro-composites by orders of magnitude with substantially lower filler loadings (often less than 5 wt.%), provided that the tribological features strongly depend on the homogeneity of dispersion and distribution of the nanofillers within the polymer matrix.

Past macro and micro models, which have been always exploited for estimating the mechanical behavior of composite materials seem to be quite inadequate to describe the phenomena occurring at a nanoscale level, particularly referring to wear and friction.

The standard tools applied for characterizing nanomaterials need to be implemented more in tribology studies to help clarify the obtained experimental results. This means that tribology should always be considered as an important issue of the materials science.

In particular, regardless of the effectiveness of the nanofiller dispersion within the polymer matrix, some issues become very crucial and should be consequently deeply investigated. First of all, the chemistry and chemical reactions, which may occur in between the mating surfaces, have to be considered, and the influence of the by-products resulting from such reactions or during wear as well.

Indeed, the effect and dynamics of the development of the transfer film during low wear sliding, together with the evolution of its physico-chemical and mechanical properties should be thoroughly investigated. Consequently, the mechanisms, through which removal of abraded materials occurs, should be deeply investigated, so that proper mechanics models for the design of high wear resistant nanocomposites can be developed.

Finally, synergies between materials science and tribology have to be developed, aiming to better understand the complex tribological phenomena taking place in polymeric nanocomposites.

This approach will surely contribute to design more efficient nanomaterials for tribological applications.

4. Acknowledgment

The financial support of Piedmont Region, Italy (Innovative Systems for Environmental friendly air COMPression - ISECOMP Project 424/09 - Piedmont Region industrial research call 2008) is gratefully acknowledged.

5. References

Avella, M.; Errico, M.E., Gentile, G. (2007). PMMA Based Nanocomposites Filled with Modified $CaCO_3$ Nanoparticles. *Macromolecular Symposia,* Vol.247, pp. 140-146

Avolio, R.; Gentile, G., Avella, M., Capitani, D., Errico, M.E. (2010). Synthesis and Characterization of Poly(methylmethacrylate)/Silica Nanocomposites: Study of the Interphase by Solid-State NMR and Structure/Properties Relationships. *Journal of Polymer Science: Part A: Polymer Chemistry,* Vol.48, pp. 5618-5629

Bahadur, S.; (2000). The Development of Transfer Layers and their Role in Polymer Tribology. *Wear,* Vol.245, pp. 92-99

Bahadur, S.; Gong, D., Anderegg, J. (1996). Investigation of the Influence of CaS, CaO and CaF2 Fillers on the Transfer and Wear of Nylon by Microscopy and XPS Analysis. *Wear,* Vol.197, pp. 271-279

Barus, S.; Zanetti, M., Lazzari, M., Costa, L. (2009). Preparation of Polymer Hybrid Nanocomposites Based on PE and Nanosilica. *Polymer,* Vol.50, pp. 2595-2600

Bhimarai, P.; Burris, D.L., Action, J., Sawyer, W.G, Toney, C.G., Siegel, R.W., Schadler, L.S. (2005). Effect of Matrix Morphology on the Wear and Friction Behavior of Alumina Nanoparticle/poly(ethylene) terephthalate Composites. *Wear,* Vol.258, pp. 1437-1443

Bhimarai, P.; Burris, D.L., Action, J., Sawyer, W.G, Toney, C.G., Siegel, R.W., Schadler, L.S. (2008). Tribological Investigation of the Effects of Particle Size, Loading and Crystallinity on Poly(ethylene) terephthalate Nanocomposites. *Wear,* Vol.264, pp. 632-637

Briscoe, B.J.; Sinha, S.K. (2002). Wear of Polymers. *Proceedings of the Institution of Mechanical Engineers, Part J: Journal of Engineering Tribology,* Vol.216, pp. 401-413

Burris, D.L.; Sawyer, W.G. (2006). Improved Wear Resistance in Alumina-PTFE Nanocomposites with Irregular Shaped Nanoparticles. *Wear,* Vol.260, pp. 915-918

Burris, D.L.; Boesl, B., Bourne, G.R., Sawyer, W.G. (2007). Polymeric Nanocomposites for Tribological Applications. *Macromolecular Materials and Engineering,* Vol.292, pp. 387-402

Carrion, F.J.; Arribas, A., Bermudez, A.K., Guillamon, A. (2008). Physical and tribological Properties of a New Polycarbonate-organoclay Nanocomposite. *European Polymer Journal ,* Vol.44, pp. 968-977

Carrion, F.J.; Espejo, C., Sanes, J., Bermudez, A.K. (2010). Single-walled Carbon Nanotubes Modified by Ionic Liquid as Antiwear Additives of Thermoplastics. *Composite Science and Tecnology,* Vol.70, pp. 2160-2167

Chang, L.; Zhang, Z., Zhang, H., Friedrich, K. (2005). Effect of Nanoparticles on the Tribological Behaviour of Short Carbon Fibre Reinforced poly(etherimide) Composites. *Tribology International,* Vol.38, pp. 966-973

Chang, L.; Zhang, Z., Zhang, H., Schlarb, A.K. (2006). On the Sliding Wear of Nanoparticle Filled Polyamide 66 Composites. *Composite Science and Technology,* Vol.66, pp. 3188-3198

Chen, W.; Li, F., Han, G., Xia, J., Wang, L., Tu, J., Xu, Z. (2003). Tribological Behavior of Carbon-nanotube-filled PTFE Composites. *Tribology Letters,* Vol.15, pp. 275-278

Czichos, H. (2001). Tribology and Its Many Facets: From Macroscopic to Microscopic and Nano-scale Phenomena. *Meccanica,* Vol.50, pp. 605-615

Dasari, A.; Yu, Z.Z., Mai, Y.K. (2009). Fundamental Aspects and Recent Progress on Wear/scratch Damage in Polymer Nanocomposites. *Materials Science and Engineering R,* Vol.63, pp. 31-80

Dasari, A.; Yu, Z.Z., Mai, Y.K., Hu G.H., Varlet, J. (2005). Clay Exfoliation and Organic Modification on Wear of Nylon 6 Nanocomposites Processed by Different Routes. *Composite Science and Technology,* Vol.65, pp. 2314-2328

Dong, B.; Wang, C., He, B.L., Li, H.L. (2008). Preparation and Tribological Properties of Poly(methyl methacrylate)/Styrene/MWNTs Copolymer Nanocomposites. *Journal of Applied Polymer Science,* Vol.108, pp. 1675-1679

Feldman, D. & Barbalata, A. (1996). *Synthetic Polymers : Technology, Properties, Applications,* Springer, New York, USA

Garcia, M.; de Rooij, M., Winnbust, L., van Zyl, W.E., Verweij, H. (2004). Friction and Wear Studies on Nylon 6/SiO_2 Nanocomposites. *Journal of Applied Polymer Science,* Vol.92, pp. 1855-1862

Hou, X.; Shan, C.X., Choy, K.W. (2008). Microstructures and Tribological Properties of PEEK-based Nanocomposites Coatings Incorporating Inorganic Fullerene-like nanoparticles. *Surface & Coatings Technology,* Vol.202, pp. 2287-2291

Johnson, B.B.; Santare, M.H., Novotny, J.E., Advani, S.G. (2009). Wear Behavior of Carbon Nanotube/High Density Polyethylene Composites. *Mechanics of Materials,* Vol.41, pp. 1108-1115

Karger-Kocsis, J.; Zhang, Z. (2005)Structure-property Relationships in Nanoparticle/semi-crystalline Thermoplastic Composites. In Balta Calleja, J.F., Michler, G., editors. In *Mechanical Properties of Polymers Based on Nanostructure and Morphology.* CRC Press, pp. 547-596, New York, USA

Kurokawa, M.; Uchiyama, Y., Nagai, S. (2000). Tribological properties and gear performance of polyoxymethylene composites. *Journal of Tribology, ASME,* Vol.122, pp. 809-814

Lee, J.Y.; Lim, D.P., Lim, D.S. (2007). Tribological Behaviour of PTFE Nanocomposite Films Reinforced with Carbon Nanoparticles. *Composites: Part B,* Vol.38, pp. 810-816

Li, F.; Hu, K., Li, J., Zhao, B. (2001). The Friction and Wear of Nanometer ZnO Filled Polytetrafluoroethylene. *Wear,* Vol.249, pp.877-882

Li, B.; Wood, W., Baker, L., Sui, G., Leer, C., Zhong, W.H. (2010). Effectual Dispersion of Carbon Nanofibers in Polyetherimide Composites and their Mechanical and Tribological Properties. *Polymer Engineering and Science,* Vol.50, pp. 1914-1922

Malucelli, G.; Palmero, P., Ronchetti, S., Delmastro, A., Montanaro, M. (2010). Effect of Various Alumina Nanofillers on the Thermal and Mechanical Behavior of Low-density polyethylene-Al_2O_3 Composites. *Polymer International,* Vol.59, pp. 1084-1089

McCook, N.L.; Hamilton, M.A., Burris, D.L., Sawyer, W.G. (2007). Tribological Results of PEEK Nanocomposites in Dry Sliding Against 440C in Various Gas Environments. *Wear,* Vol.262, pp. 1511-1515

Misra, R.; Fu, X.B., Morgan, S.E. (2006). Surface Energetics, Dispersion, and Nanotribomechanical Behavior of POSS/PP Hybrid Nanocomposites. *Journal of Polymer Science: Part B: Polymer Physics,* Vol.45, pp. 2441-2455, DOI 10.1002/polb

Peng, Q.Y.; Cong, P.H., Liu, T.X., Huang, S., Li, T.S. (2009). The Preparation of PVDF/clay Nanocomposites and the Investigation of their Tribological Properties. *Wear,* Vol.266, pp. 713-720

Ravi Kumar, B.N.; Suresha, B., Venkataramareddy M. (2009). Effect of Particulate Fillers on Mechanical and Abrasive Wear Behaviour of Polyamide 66/polypropylene nanocomposites. *Materials and Design*, Vol.30, pp. 3852-3858

Sawyer, W.G.; Freudenberg, K., Bhimaraj, P., Schadler, L. (2003). A Study on the Friction and Wear Behavior of PTFE Filled with Alumina Nanoparticles. *Wear*, Vol.254, pp. 573-580

Shi, Y.; Feng, X., Wang, H., Lu, X., Shen, J. (2007). Tribological and Mechanical Properties of Carbon-Nanofiber-Filled Polytetrafluoroethylene Composites. *Journal of Applied Polymer Science,* Vol.104, pp. 2430-2437, DOI 10.1002/app.23951

Sirong, Y.; Zhongzhen, Y., Yiu-Wing M. (2007). Effects of SEBS-g-MA on Tribological Behavior of Nylon 66/organoclay Nanocomposites. *Tribology International,* Vol.40, pp. 855-862

Sun, L.H.; Yang, Z.G., Li, X.H. (2008). Mechanical and Tribological Properties of Polyoxymethylene Modified with Nanoparticles and Solid Lubricants. *Polymer Engineering and Science,* Vol.48, pp. 1824-1832

Sun, L.H.; Yang, Z.G., Li, X.H. (2008). Study on the Friction and Wear Behavior of POM/Al_2O_3 Nanocomposites. *Wear*, Vol.264, pp. 693-700

Vail, J.R.; Burris, D.L., Sawyer, W.G. (2009). Multifunctionality of Single-walled Carbon Nanotube-polytetrafluoroethylene nanocomposites. *Wear*, Vol.267, pp. 619-624

Wang, Y.; Hu, X.G., Tian, M., Stengler, R. (2007). Study on Mechanical and Tribological Property of Nanometer ZrO_2-filled Polyoxymethylene Composites. Polymer-Plastics Technology and Engineering, *Vol.46, pp. 469-473*

Wang, Y.; Hu, K.H., Xu, Y.F., Hu, X.G. (2008). Structural, Thermal, and Tribological Properties of Intercalated Polyoxymethylene/molybdenum Disulfide Nanocomposites. *Journal of Applied Polymer Science,* Vol.110, pp. 91-96

Wang, Y.; Lim, S., Luo, J.L., Zu, Z.H. (2006). Tribological and Corrosion Behaviour of Al_2O_3/polymer Nanocomposite Coatings. *Wear,* Vol.260, pp. 976-983

Werner, P.; Altstadt, V., Jaskula, R., Jacobs, O., Sandler, J.K.W., Shaffer, M.S.P., Windle, A.H. (2004). Tribological Behaviour of Carbon-nanofibre-reinforced Poly(ether ether ketone). *Wear,* Vol.257, pp. 1006-1014

Xian, G.; Zhang, Z., Friedrich, K. (2006). Tribological Properties of Micro- and Nanoparticles-Filled Poly(etherimide) Composites. *Journal of Applied Polymer Science,* Vol.101, pp. 1678-1686

Xiong, L.; Xiong, D., Yang, Y., Jin, J. (2011). Friction, Wear, and Tensile properties of vacuum hot pressing crosslinked UHMWPE/nano-HAP composites. *Journal of Biomedical Materials Research B: Applied Biomaterials,* Vol.98B, issue 1, pp. 127-138

Zhang, G. (2010). Structure-Tribological Property Relationship of Nanoparticles and Short Carbon Fibers Reinforced PEEK Hybrid Composites. *Journal of Polymer Science: Part B: Polymer Physics,* Vol.48, pp. 801-811

Zhang, G.; Chang, L., Schlarb, A.K. (2009). The roles of Nano-SiO_2 particles on the tribological behaviour of short carbon fiber reinforced PEEK. *Composite Science and Technology,* Vol.69, pp. 1029-1035

Zhang, G.; Schlarb, A.K., Tria, S., Elkedim, O. (2008). Tensile and Tribological Behaviors of PEEK/nano-SiO_2 Composites Compounded Using a Ball Milling Technique. *Composite Science and Technology,* Vol.68, pp. 3073-3080

Zhang, H.; Zhang, Z., Guo, F., Wang, K., Jiang, W. (2009). Enhanced Wear Properties of Hybrid PTFE/cotton Fabric Composites Filled with Functionalized Multi-walled Carbon Nanotubes. *Materials Chemistry and Physics,* Vol.116, pp. 183-190

Zhong, Y.J.; Xie, G.Y., Sui, G.X., Yang, R. (2011). Poly(ether ether ketone) Composites Reinforced by Short Carbon Fibers and Zirconium Dioxide Nanoparticles: Mechanical Properties and Sliding Wear Behavior with Water Lubrication. *Journal of Applied Polymer Science,* Vol.119, pp. 1711-1720

Zhou, Q.; Wang, K., Loo, L.S. (2009). Abrasion Studies of Nylon 6/Montmorillonite Nanocomposites Using Scanning Electron Microscopy, Fourier Transform Infrared Spectroscopy, and X-ray Photoelectron Spectroscopy. *Journal of Applied Polymer Science,* Vol.113, pp. 3286-3293

2

Abrasive Effects Observed in Concrete Hydraulic Surfaces of Dams and Application of Repair Materials

José Carlos Alves Galvão,
Kleber Franke Portella and Aline Christiane Morales Kormann
Federal Technological University of Paraná, Institute of Technology for Development, Federal University of Paraná
Brazil

1. Introduction

This chapter presents the main abrasive effects observed in concrete hydraulic surfaces of dams of hydroelectric power plant (HPP). These types of hydraulic structures are subject to solicitations for dynamics order due to water flow at high speed usually causing to erosion, cavitation and abrasion. These effects are harmful and cause of defects on the surface hydraulic of dam. Thus, maintenance is needed on the surface of the structure, when is applied a repair material (RM). The RMs must have characteristics, especially mechanical and chemical properties, consistent with the base material or substrate.

We analyzed the defects from the abrasive processes caused by the flow of water from the reservoir that occurred in two concrete dams in Brazil. One is located in the state of São Paulo, southeastern Brazil. The other is located in the state of Paraná, southern of country. Have been proposed RM different for application at surfaces hydraulic of dams, which had their performance verified in laboratory and field.

The steel fiber concrete was developed based on the concrete mixture used in one slab of the spillway for the dam in the Southeast, case one. The average mixture used was 1: 1.61: 2.99: 0.376, with cement consumption of 425 kg/m^3. The mortar mixtures were made following the information given by the manufacturers. Tests of the RM were conducted in laboratory: abrasion resistance of concrete (underwater method), flexural tensile resistance, compressive strength tests, elasticity modulus, resistance to adherence, accelerated aging in UV ray chamber and humidity and permeability tests.

In case study 2 were analyzed in the laboratory and applied field, the dam of the South, were made from mixtures of concrete with addition polymeric and elastomer materials proceeding from the recycling industry, such as agglutinated low-density polyethylene (LDPE), crushed polyethylene terephthalate (PET) and rubber from useless tires. The contents of recycled material were 0.5%, 1.0%, 2.5%, 5.0% and 7.5%. Mixtures with added recycled material and RC were analyzed in the laboratory in the mechanical properties of compressive strength, splitting tensile strength and grip. Considering the results of the laboratory were selected for field application the concretes with contents of 2.5 and 5.0%. Besides the concretes with addition of recycled materials were applied in field. The materials used in the field were examined in tests of abrasion resistance of concrete (underwater method) and accelerated aging tests performed in a moist chamber with SO_2 and baths by immersion in NaCl, Na_2SO_4 and distilled water, where the specimens were followed by the technique of the corrosion potential.

The tests were carried out in laboratory on concrete samples in order to simulate the environmental conditions, which are usually found, among others, for controlling the mechanical resistance and the aging imposed conditions, such as solar radiation, humidity and chemical attack.

2. Mechanisms of surface wear in hydraulic structures of concrete

The durability of a concrete structure is strongly influenced by the inappropriate use of materials and physical and chemical effects of the environment where it operates. The immediate consequence is the anticipated need of maintenance and execution of repairs (Galvão et al, 2011).

In the case of hydraulic structures of concrete, one of the main forms of degradation is related to abrasive processes. In general, the erosion caused by surface wear of the hydraulic structures of concrete, is defined as the disintegration of the material exposed to the phenomena of deterioration (Kormann, 2002).

Normally, concrete is measured and produced by following certain criteria for structural and operational conditions that can support the loads and overloads for several years without wear. However, for a variety of factors, including design parameters and construction, selection and quality of materials, operational changes, as well as interaction with the environment, the structures are damaged, and the degree of deterioration is directly related to these factors.

To recover the surfaces that have suffered such damage, various materials and application techniques have been developed. These repair materials should be appropriate to the characteristics of the phenomenon of wear as well as the operating conditions of the structures. Other considerations, such as access to sites of repair, time of execution of services, cost of operations, staff expertise on the handling of materials and equipment, should be estimated so taht whole recovery program is carried out with full success.

Maintenance of structures on surfaces of concrete dams should be done by combining the characteristics of cost, feasibility, performance, durability, usage, time of application of the materials and compatibility between them (Kormann et al, 2003).

2.1 Physical causes of concrete deterioration

Physical causes of concrete deterioration were grouped by Metha & Gerwick (1992) in two categories. The first category consists of the surface wear due to abrasion, erosion and cavitation. The second category is composed by cracking due to volume variation, structural loading and exposure to high temperatures.

According to Mehta & Monteiro (2006), the term abrasion refers to dry friction, as in the case of wear of industrial floors and pavements due to vehicle traffic. Erosion is usually used to describe the wear by the abrasive action of fluids satisfaction suspended solids occurring as coatings on hydraulic structures of channels and spillways. The cavitation damage to hydraulic structures, and relates to the loss of mass by the formation of vapor bubbles caused sudden change of direction in rapidly flowing waters.

In the manual of the American Concrete Institute (1999) are considered like erosive processes in concrete structures: cavitation, abrasion and wear by chemical attack.

This differentiation of works cited is purely arbitrary, as emphasized by Mehta & Monteiro (2006). Generally, the physical and chemical damage, ultimately complement each other. When occur physical damage, such as abrasion, there is increased exposure of the concrete surface to agents such as acid rain, and therefore the attack by chemical compounds is favoured. When occur the chemical damage, such as leaching, the concrete is more porous, facilitating the process of abrasion, and so on. These facts make both processes of deterioration, physical or chemical, a cycle of difficult to dissociation or stabilization.

It is understood then, that the abrasion term refers to wear by dry friction and the erosion term is the wear by the impact of suspended solids carried by a fluid (Neville, 1996). In hydraulic structures for dams this fluid is water.

Although the terms abrasion and erosion differ by the type of environment in which the wear occurs, dry or suspended in water, wear that occurred on concrete surfaces hydraulic is called erosion by abrasion or simply abrasion.

2.2 Occurrence of abrasion in concrete hydraulic structures

Abrasion is caused by the impact of elements transported by water in hydraulic structures of concrete. How much more turbulent are the flows, along with the impact forces caused by debris, the greater the abrasion.

The debris transported by water flows ranging from their hardness until their types, and can be sand, stones, rubble, gravel, etc. The hydraulic structures most affected by abrasive processes are the surfaces of the spillways, stilling basin, walls of the upstream reservoir, drain pipes and hydraulic tunnels.

In hydraulic structures of concrete dams, turbulent flows of water with suspended debris, colliding into their concrete surfaces, can cause abrasions to various depths. Great damage by abrasion occurred at Dworshak Dam, whose abrasion consumed an approximate volume of concrete and foundation rock of 1,530 m^3, and approximate depths of 2 and 3 m (ACI, 1999)

2.3 The main factors affecting the resistance of concrete abrasion

The main factors affecting the abrasion resistance of concrete are the environmental conditions and dosing of aggregates, concrete strength, the mix ratio, the use of special cement, the use of supplementary materials, such as adding fiber and fly ash . Two other factors have an important effect on the abrasion resistance, surface finish and curing conditions (Horszczaruk, 2005).

The compressive strength proved to be one of the most important factors that correlate with the abrasion resistance of concrete. The compressive strength does not influence the abrasion resistance however is verified a correlation between them, if one is high; the other tends to be too.

Holland et al. (1987) established dependence between the concrete compressive strength and abrasion resistance underwater method in 72 hours. The tests showed that the abrasion resistance increases with the compressive strength. Holland studied the abrasion resistance of concrete with 11 to 15% silica fume and water/cement ratio (w/c) ranging between 0.24 and 0.34 to repair the Kinzua Dam in Pennsylvania. The concrete had, after 28 days, compressive strength of until 79MPa. The use of silica fume improved abrasion resistance compared to conventional concrete.

In the work of Horszczaruk (2005) presents the program takes nine high-strength concrete made of various types of cement and modified with steel fiber, PVC fiber and latex. The concrete was made with three types of cement: Portland CEM I 42.5R, CEM I 52.5R and CEM III 42.5. All mixtures contained fly ash (SiO_2, 93%) and silica fume (10% of cement mass). The mixtures contained the fraction of basalt aggregate with density of 3.03 kg/m3 and maximum size of 8 and 16 mm. The paper presents important findings: i) The ASTM C1138 method is suitable for determining the abrasion resistance of the high-strength concrete (HSC) to 28 days compressive strength of 80 to 120 MPa; ii) The period analysis method for underwater (HSC) should be at least 72 h; iii) The assumption of linear dependence of wear (HSC) is correct, omit the first stage of abrasion (12-24h). The rate of wear can be assumed constant (HSC over 80 MPa); and iv) The latex additive does not improve the abrasion resistance of concrete. The HSC with added PVC fiber showed improvement in this area.

2.4 Repair of concrete hydraulic structures

Repairs to damaged concrete structures are important not only to ensure the planned useful life, but also to provide good performance and security facing the most severe applications.

An adequate repair improves the function and performance of the structure, restores and increases its strength and stiffness, improves the appearance of the concrete surface, provides impermeability to water, prevents the penetration of aggressive species at the interface concrete/steel and improves its durability (Al-Zahrani et al, 2003).

The surfaces of hydraulic concrete dams are subject to wear by erosion (American Concrete Institute, 1999), fissures caused by the pressure of crystallization of salts in the pores (Tambelli et al, 2006) and by exposure to contaminants (Irassar et al, 2003), causing defects and constant maintenance and repair applications.

Various materials are marketed for repair of deteriorated concrete structures. The RM is the most commonly used is the mortars with silica fume, epoxy resin and polyester resin, the concrete with polymers and concrete reinforced with fibers.

To carry out repairs to a concrete structure should be considered the main causes of defects, the extent of deterioration, environmental conditions and external stresses imposed. Since then, it follows by the choice of RM itself, which meets the design specifications, as schematic design presented in Fig. 1.

The recovery services of hydraulic structures are extremely expensive. According to Smoak (1998), recovery operations and maintenance of infrastructure of water resources, located mainly in the most severe climatic zones of the United States accounted for spending more than U$ 17 billion.

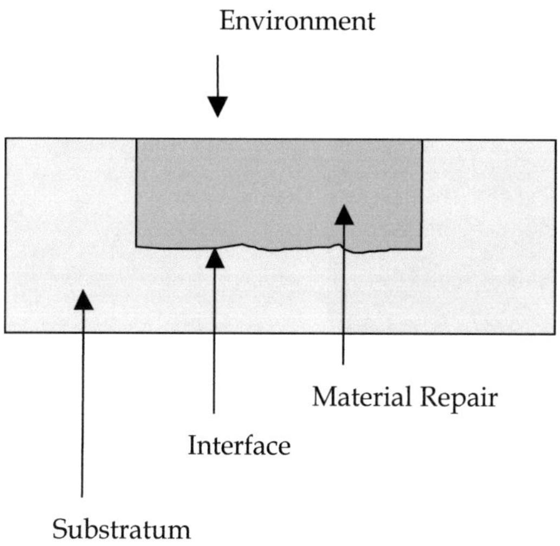

Fig. 1. Repair system: substrate/material repair

In Brazil, the concern about the safety of dams usually relates to the structural problems, mainly due to the catastrophic results of the loss of generation, supply capacity, indemnity expenses, cost of recovery and depreciation in the name and prestige of the Company (Cardia, 2008).

3. Case studies for the application of repair materials

We evaluated the state of degradation and the need to repair the spillway of two hydroelectric power plants, located in the south and southeastern of the Brazil. The hydraulic structures of these HPP have been degraded by the processes of abrasion, and thus, different materials were applied concrete repair these dams.

3.1 Case one

When performed visual inspection of the dam in southeastern Brazil were analyzed various structures such as the surface of the spillway surface of the slab, side walls, pillars of the gates and blocks of dissipation. These structures were found several points of deterioration and abrasive processes.

As noted in the inspections, the state of degradation of the dam due to abrasion processes requires additional repairs to the concrete hydraulic structure. To repair the dam shown in Fig, were proposed four types of RM: mortar with silica fume, epoxy mortar, mortar and polymer concrete with steel fibers. These repair materials were evaluated for mechanical properties of strength and durability. The highlight is the test of abrasion resistance according to ASTM C 1138. The samples subjected to abrasion tests were evaluated according to the wear surface.

Is shown in Fig. 2a, overview of the spillway. It is observed that the wall of the gate shows signs of erosion (highlighted). Is shown in Fig 2b, defect in a slab of the spillway caused by abrasion.

(a) (b)

Fig. 2. (a) Spillway dam presented deterioration by abrasion. (b) Defects in the concrete structure caused by abrasive process

3.1.1 Repair materials

a. Mortar with silica fume

According Ghafoor & Diawara (1999), the optimum silica fume content is around 10% mass of cement, replacing the fine aggregate. The proportion of the mixture used to construct the mortar with silica fume was 1: 3.66: 0.5: 0.1 (cement: sand: water: silica).

b. Epoxy mortar

The epoxy mortar was measured according to manufacturer's instructions. This type of material is composed of components A (resin) + B (hardener) and C (quartz sand).

c. Polymer mortar

For the production of polymer mortar was used an industrial product in which it was necessary to add only water. The water content indicated by the manufacturer was around 18%. However, the best workability was checked with a water content of 11%, which was then used in making the RM.

d. Concrete with steel fiber

The mixture used to prepare the RM of concrete with addition of steel fibers was 1: 1.985: 2.77: 0.45: 2.42 (cement: sand: gravel: water: steel fibers).

e. Concrete Reference

Although the mixture was made of the material considered as the reference concrete (RC) with characteristics similar to concrete used in dam construction. The content of the mixture of the RC was 1: 1.61: 2.99: 0.376 (cement: sand: gravel: water) with cement content of 426 kg/m^3.

3.1.2 Compressive strength

The compressive strength is considered one of the main evaluation parameters for resistance of cementitious materials on the request abrasion (Mehta & Monteiro, 2006; American

Concrete Institute, 1999; Neville, 1996). The values were analyzed for characteristic strength at 28 days of curing material. The average results were 48.3 MPa for the reference concrete, 45.4 MPa for the mortar with silica fume, 85.4 MPa for the epoxy mortar, 40.7 MPa for the polymer mortar and 39.7 MPa for concrete with steel fibers.

3.1.3 Preparation of specimen for testing abrasion underwater method

For the analysis of resistance to abrasion submerged by the method followed ASTM C 1138/97. Were fabricated specimens of concrete with diameters of 300 mm and height of 100 mm. On the test specimens was a void top center of approximate diameter of 200 mm and 50 mm height. The body of evidence resulting from this process was considered the substrate structure. After his cure at 28 days, resulting in gaps of various specimens were filled with the RM study. The preparation procedure of specimens is illustrated in Fig. 3. After 28 days of curing of the specimen substrate + RM, in a wet chamber with relative humidity greater than 95% and controlled temperature (23 ± 2) ° C, we evaluated the abrasion resistance of the repair systems studied.

(a) (b)

Fig. 3. Preparation of specimens for testing abrasion. (a) Release and compacting concrete reference. (b) System: substrate | repair material.

3.1.4 Abrasion resistance of concrete

In Table 1, the values of mass loss, after 72h of abrasion (underwater method), of RC and system CR | RM, are presented.

Repair material	mass loss (%)		
	24 h	48 h	72 h
Mortar with silica fume	3.33	4.80	5.53
Epoxy mortar	1.31	1.66	2.03
Polymer mortar	4.72	5.97	6.40
Concrete with steel fiber	1.57	2.96	3.64

Table 1. Mass loss of concrete specimens under abrasion (underwater method).

The average mass loss for the substrate system | mortar with silica fume was 5.53% and was virtually in the region of the substrate and the interface between both materials.

Fig. 4, shows the performance of the system substrate | mortar with silica fume, 72 hours after the test.

(a) (b)

Fig. 4. Results of testing of abrasion resistance of the RM with silica fume. (a) Specimen after test. (b) 3D schematic graph of wear occurred in the abrasion resistance test (underwater method).

The mass loss for the substrate system | epoxy mortar was averaging 2%. It can be seen in Fig. 5 that much of the material was extracted from the substrate, with little influence in the region of interface. This effect can be attributed to high compressive strength of the material (over 90 MPa).

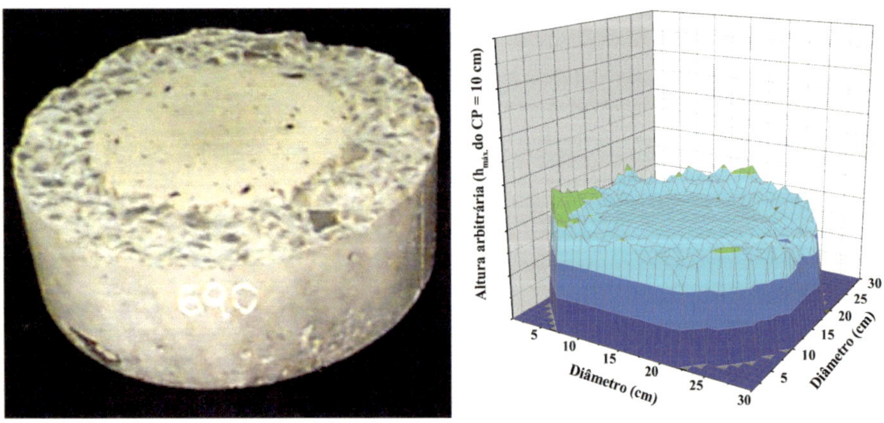

(a) (b)

Fig. 5. Results of testing of abrasion resistance of the RM with epoxy mortar .
(a) Specimen after test. (b) 3D schematic graph of wear occurred in the abrasion resistance test (underwater method).

In Fig. 6, shows the performance of the system substrate | polymer mortar. The average weight loss was 6.4% and located in the substrate interface | polymer mortar.

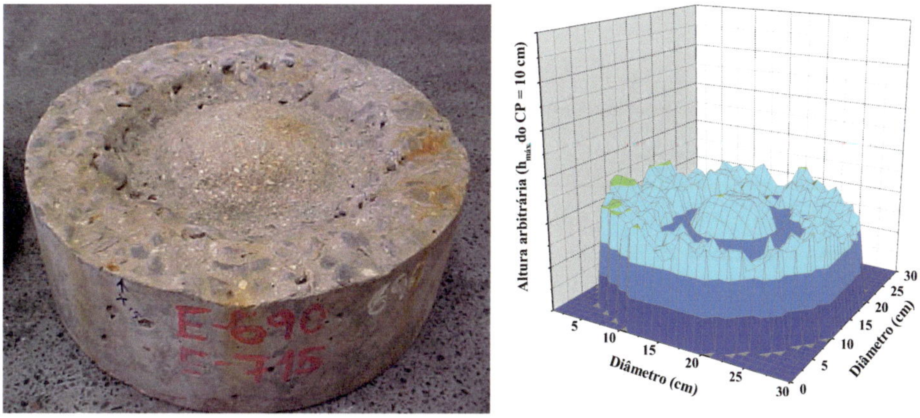

(a) (b)

Fig. 6. Results of testing of abrasion resistance of the RM with polymer mortar . (a) Specimen after test. (b) 3D schematic graph of wear occurred in the abrasion resistance test (underwater method).

The effect of abrasion on the test substrate system | concrete with steel fibers was also not very pronounced. The mass loss was more intense in the region of RC, as shown in Fig. 7.

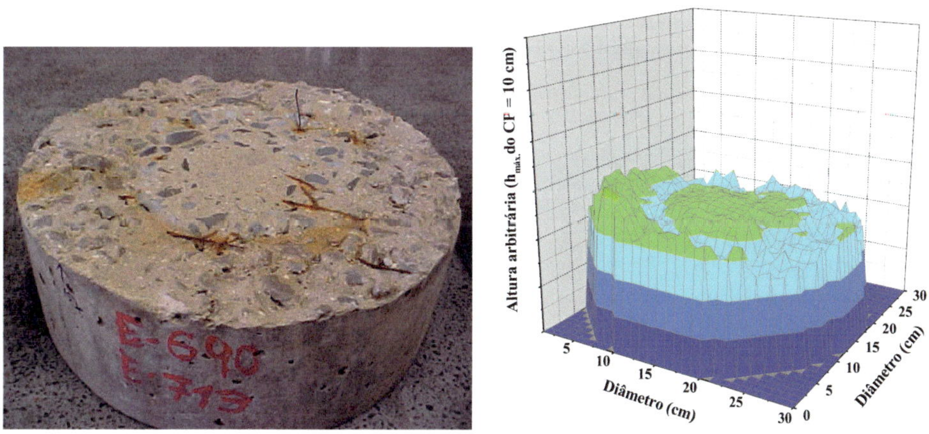

(a) (b)

Fig. 7. Results of testing of abrasion resistance of the RM with concrete with steel fibers. (a) Specimen after test. (b) 3D schematic graph of wear occurred in the abrasion resistance test (underwater method).

Horszczaruk (2009) conducted a study with high performance concrete (HPC) and high performance fiber-reinforced concrete (HPFCR) to analyze performance and hydraulic abrasion. The author found that the effect called the shadow zone. This effect was observed specifically in concrete reinforced with steel fibers than in the polymer concrete reinforced with polypropylene fibers. The effect of shadow zone caused a slight increase in abrasion resistance of HPC.

As shown in the graphs of Fig. 4 - 7, may be the poorest performance of the mortars with silica fume and polymer mortar, respectively, the second has more wear. The substrate system | concrete with steel fibers showed a good performance and wear observed was widespread. Already, for the substrate system | epoxy mortar, the wear was observed on the substrate with little or no loss of epoxy mortar.

The information obtained in the 3D schematic drawings of wear occurred in the abrasion resistance test underwater method corroborate the results in Table 1, where the materials of greater mass loss were made with RM mortar with silica fume and polymer mortar. Concrete reinforced with steel fibers had lower mass loss that these two mortars possibly due to the effect of shadow zone. The epoxy mortar, material with higher mechanical strength (85.4 MPa), showed little wear (Fig. 5) and therefore represented by the lower mass loss during the 72 hour test (Table 1).

3.2 Case two

As a case study 2 assessed the repair of the spillway chute Hydroelectric Plant located in southern Brazil, with the use of concrete repair materials with the addition of polymeric materials. Seeking a proposed sustainable polymeric materials were analyzed from the recycling industry. The materials agglutinated low-density polyethylene (LDPE), crushed polyethylene terephthalate (PET) and rubber from useless tires were added to the concrete for the construction of the repair material degrades the surface hydraulic and abrasive processes.

In Fig. 8a presents the overview of the spillway of the dam and in Fig. 8b, a detail of the displacement of the surface caused by hydraulic abrasive processes.

(a) (b)

Fig. 8. (a) Overview of the spillway of the dam. (b) Detail of the displacement of the surface caused by hydraulic abrasive processes.

3.2.1 Compressive strength

To determine the best mix for making RM were added to the concrete, recycled polymeric material in the contents of 0.5%, 1.0%, 2.5%, 5.0% and 7.5%

In Figure 9, shows the average results of compressive strength at 28 days to cure the concrete with the addition of recycled polymer materials (Galvão et al, 2011).

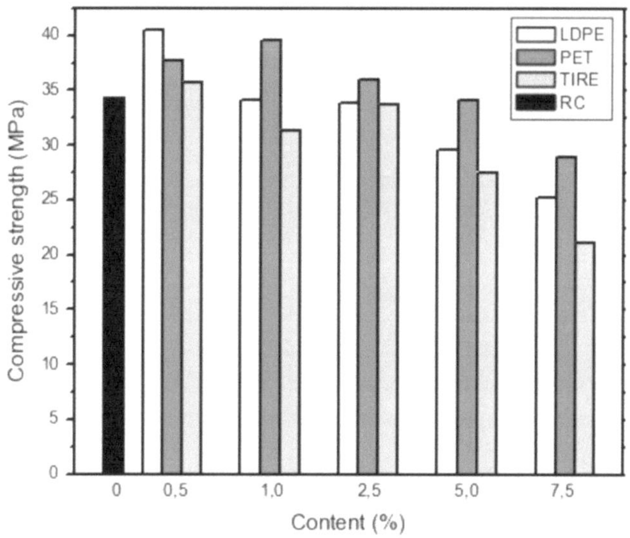

Fig. 9. Comparative graph of compressive strength, at 28 days, of concretes with additions of waste and its respective contents.

For the different materials studied and their respective contents verified that the results obtained from tests for compressive strength as a function of curing time were inversely proportional to the content of addition of recycled fibers. This trend was also observed by Choi et al (2005) when they studied the effects of adding waste PET bottles in the properties of concrete and by Al-Ismail & Hashmi (2008) in the bond strength between the surface of the waste polymer and cement paste.

The reduction in compressive strength of concrete mixtures with the addition of plastic aggregates may be due to a lower resistance of these particles, when compared to natural aggregate (Batayneh et al, 2007).

Al-Manas and Dalal (1997) and Soroushian et al (2003) also found that the compressive strength decreased with increasing aggregate content in plastics.

In the studies by Freitas et al (2009) was verified a reduction in compressive strength with increasing rubber content of added sugar. Li et al (2004) also observed reduction in axial compressive strength of concrete with fibers tire and is not related to any such loss characteristic of the material.

In the analysis of all the values of compressive strength (Fig. 9), were chosen levels of 2.5% and 5.0% addition of recycled material for the manufacture of polymeric materials to repair the degraded surface at the dam. These were not necessarily the levels with greater resistance to axial compression. However, we considered the possibility of adding the larger volume waste polymer concrete giving them the appropriate final destination, the contents of 2.5% and 5.0% was therefore selected.

3.2.2 Abrasion resistance of concrete

The Table 2, shows the values of mass loss by abrasion of the RC and of the concretes with addition with recycled polymer materials.

Repair material	mass loss (%)		
	24 h	48 h	72 h
RC	2.50	4.24	6.57
Tire 2.5	3.26	5.22	8.41
Tire 5	3.19	5.03	7.39
PET 2.5	3.37	5.57	8.07
PET 5	2.86	4.55	5.04
LDPE 2.5	3.44	5.39	7.46
LDPE 5	1.95	3.32	3.94

Table 2. Mass loss of concrete specimens under abrasion (underwater method).

Analyzing the mass loss at 72 hours the test of resistance to abrasion (Table 2), it is observed that the CR had the best performance when compared to concrete with addition of residues in the polymer content of 2.5% . However the content of 5% waste polymer concretes with the addition of PET and LDPE showed better results than plain concrete. There is no linearity in the mass loss for 24, 48 and 72 hours of testing. Horszczaruk (2005) considers that the period of analysis method for abrasion resistance under water must be at least 72 hours for high strength concrete in hydraulic structures.

Soroushian et al (2003) observed that the addition of recycled plastic caused a reduction in abrasion resistance of concrete. The authors attributed this effect to the fact that the fibers were torn near the surface under the effect of change in the abrasion and wear characteristics of concrete with the presence of fibers from recycled plastic modified the surface characteristics of the material.

Analyzing the content of adding recycled polymeric material notes that for a greater amount of material added to the concrete there is a gain in resistance to abrasion. This increase occurs for all materials analyzed the levels of 2.5 and 5.0%. A similar effect was verified by Soroushian et al (2003) in abrasion resistance depending on the type of plastic fiber only to an increase in the content of virgin polypropylene fibers of 0.075% to 0.15% compared to the volume of cement.

Considering the type of recycled material used in addition, it appears that the concrete produced with waste LDPE have been the best performance for abrasion resistance. The material made from fibers tire had the highest values of mass loss, representing less resistance to abrasion. The concrete with the addition of PET showed intermediate values as between the other two types of waste studied.

3.2.3 Application in field

In order to check the performance of materials for the repair of hydraulic structures of concrete, there were applications of these materials studied in the spillways of HPP. RM were applied to concrete with the addition of recycled polymer materials (LDPE, PET and tire) in levels of 2.5% and 5.0%, in addition to mortar with silica fume, epoxy mortar, polymer mortar and concrete with steel fiber.

The application of RM consisted the steps of selecting points of application, cutting and scarification, dosage and application of the concrete itself. These steps are described in the work of Galvão et al (2011) and Kormann et al (2003).

The points of application of the repair materials were selected in the spillway after complete observation of the degradation state of the dam and location of defects exposed in concrete blocks.

The curved regions were avoided in the spillway for the implementation of repair material due to the existence of differential effects of cavitation, which could compromise comparative testing of field.

As a bridge bonding was applied on the dry surface substrate, a structural adhesive epoxy resin of high viscosity (Fig. 10a). This care has been taken to ensure good bond strength between the substrate and RM.

The concrete was cast into the point of application of the substrate with a trowel and pestle, removing air bubbles from the material (Fig. 10b).

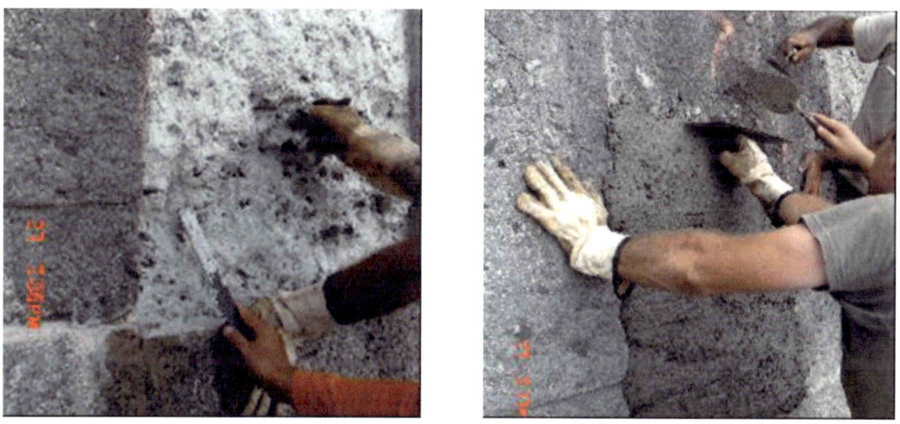

(a) (b)

Fig. 10. (a) Application of bridge bonding. (b) Launch of the repair material.

3.2.4 Performance of repair materials applied field

After two years of application, we observed the general state of conservation of repair materials, mainly related to the appearance of defects and interface between the RM and the substrate. In Fig. 11 and 12 are registered examples of points where they were applied in the field analyzed RM.

It can be seen in Fig. 11a the repair material applied to the surface of the hydraulic concrete structure is well adhered to the substrate. In the detail of Fig. 11b the interface between old concrete and the RM applied not showed flaws or defects, even after the period of water flow in the spillway of the dam and the middle attacks.

(a) (b)

Fig. 11. Application point of repair material. (a) Reference concrete without additions. (b) Detail of the border between the RM and the substrate.

(a) (b)

Fig. 12. Application point of repair material. (a) Concrete with addition of recycled LDPE in the content of 2.5%. (b) Detail of the border between the RM and the substrate.

It appears that the point of application above a superior performance compared with other repair materials. At this point the surface has good finish, smooth and free of imperfections, even after the occurrence of water flow in the spillway and aggression in the middle. At no point were visually observed cracks or fissures.

The defects in the substrate were completely contained at this point of application of RM made of concrete with added content in 2.5% recycled LDPE material.

4. Conclusion

For two case studies were analyzed in total 11 types of repair materials for hydraulic concrete surfaces.

In the case of mortars was proved the relationship between abrasion resistance and compressive strength of these materials. The repair material made of epoxy mortar performed better on the compressive strength presented as the material more resistant to abrasion underwater method. In turn, the polymer mortar material was lower mechanical strength and consequently the repair material with greater mass loss in abrasion tests.

The epoxy mortar has high mechanical strength. It was found that when the mechanical strength of the repair material was far superior to the concrete substrate, caused edge effect. Thus, the hydraulic structure, the water moving at high speed, may form a step that will facilitate the process of cavitation. The specimens with epoxy mortar in the abrasion test showed the edge effect mentioned. Was considered appropriate to seek similar mechanical strength between the substrate and repair material so that the structure suffers wear a uniform.

When we evaluated the addition of concrete with recycled polymer materials was observed that the inclusion of these materials reduced the compressive strength of concrete. The greater the content used was the lowest average compressive strength. This reduction was identified in the waste polymer when added to concrete in the contents of 0.5% to 7.5%. However, evaluating the abrasion resistance of the underwater method, considering the mass loss, it was found that the RC had the best performance when compared to concrete with addition of residues in the polymer content of 2.5%. However, the content of 5% waste polymer added to the concrete showed better results than plain concrete, indicating that, for polymer fibers, the increase in the level of addition may influence the increased resistance to abrasion

5. Acknowledgment

To Federal Technological University of Paraná; to Institute of Technology for Development by the financing and infrastructure for the conduction of works.

6. References

Al-Manaseer, A.A. & Dalal, T.R. (1997). Concrete containing plastic aggregates. *Concrete International*. Vol. 19, No. 9, (August 1997), pp. 47-52, 0162-4075

Al-Zahrani,M. M.; Maslehuddin; M, Al-Dulaijan; S. U. & Ibrahim, M. (2003). Mechanical properties and durability characteristics of polymer- and cement-based repair materials. *Cement and Concrete Composites*. Vol. 25, No. 4-5, (May-July 2003), pp. 527-537, 0008-8846

American Concrete Institute. (1999). *Concrete Repair Manual*, Document nº: ACI210.1R-94. Compendium of case histories on repair of erosion-damaged concrete in hydraulic structures

American Society for Testing and Materials, ASTM C 1138-97. (2002). *Standard test method for abrasion resistance of concrete* (underwater method), in: Annual Book of ASTM Standards, vol. 04.02, ASTM, West Conshohocken

Batayneh, M.; Marie, I. & Asi, I. (2007). Use of selected waste materials in concrete mixes. *Waste Management*. Vol. 27 , p. 1870 - 1876

Cardia, R. J. R. Auditoria em segurança e controle de barragens. *Proceedings of VI Simpósio Brasileiro Sobre Pequenas e Médias Centrais Hidrelétricas*, p. 24-39, Belo Horizonte, Minas Gerais, Brazil, Apri 21-25, 2008 (in Portuguese)

Choi, Yun-Wang, Moon, Dae-Joong, Chung, Jee-Seung & Cho, Sun-Kyu. (2005). Effects of waste PET bottles aggregate on the properties of concrete. *Cement and Concrete Research*. v. 35, p. 776-781

Freitas, C.; Galvão, J.C. A.; Portella, K. F.; Joukoski, A.; Gomes Filho, C. V. & Ferreira, E. S. (2009). Desempenho físico-químico e mecânico de concreto de cimento Portland com borracha de estireno-butadieno reciclada de pneus. *Química Nova*. v. 32, p. 913-918

Galvão, J. C. A.; Portella, K. F.; Joukoski, A.; Mendes, R. & Ferreira, E. S. (2011). Use of waste polymers in concrete for repair of dam hydraulic surfaces. *Construction and Building Materials*. v. 25, p. 1049–1055

Ghafoori, N. & Diawara, H. (1999). Abrasion resistanceof fine aggregatereplacedsilica fume concrete. *ACI Materials Journal* , v. 96, p. 555-567

Holland, T.C. (1987). Erosion resistance with silica-fume concrete. *ACI Materials Journal*. p. 32–40

Horszczaruk, E. (2005). Abrasion resistance of high-strength concrete in hydraulic structures. *Wear*. v. 259, p 62-69

Horszczaruk, E. (2009). Hydro-abrasive erosion of high performance fiber-reinforced concrete. *Wear*. v. 267, p. 110-115

Irassar, E. F.; Bonavetti, V.L. & Gonza´Lez, M. (2003). Microstructural study of sulfate attack on ordinary and limestone Portland cements at ambient temperature. *Cement and Concrete Research*. v. 33, p.31-41

Ismail Z. Z. & Al-Hashmi, E. A. (2008). Use of waste plastic in concrete mixture as aggregate replacement. *Waste Management*. v. 28, p. 2041-2047

Li, G.; Garrick, G.; Eggers, J.; Abadie, C.; Stubblefield, M. A. & Pang, S. (2004). Waste tire fiber modified concrete. *Composites: Part B engineering*. v. 35 p. 305–312

Kormann, A.C.M. (2002). *Estudo do desempenho de quatro tipos de materiais para reparo a serem utilizados em superfícies erodidas de concreto de barragens,* M.Sc, Dissertation, UFPR, Curitiba, Brazil

Kormann, A.C.M., Portella, K. F.; Pereira, P. N. & Santos, R. P. (2003). Study of the performance of four repairing material systems for hydraulic structures of concrete dams. *Cerâmica*. v. 49, p. 48-54

Metha, P. K. & Gerwick Jr, B. C. (1982). Cracking-Corrosion Interaction in Concrete Exposed to Marine Environment. *Concrete International*. v. 10, p. 45-51

Mehta, P. K. & Monteiro, P. J. M. (2006). *Concrete: microstructure, properties and materials,* McGraw-Hill, Columbus, USA

Neville, A. M. *Properties of Concrete,* 4th ed., John Wiley & Sons, Inc., New York, 1996, USA

Smoak W. G. (1998). *Guide to Concrete Repair,* Reclamation Bureau, Washington DC, USA

Soroushian, P.; Plasencia, J. & Ravanbakhsh, S. (2003). Assessment of reinforcing effects of recycled plastic and paper in concrete. *ACI Materials Journal*. v. 100, p. 203-207

Tambelli, C.E.; Schneider, J.F.; Hasparyk, N.P. & Monteiro, P.J.M. (2006). Study of the structure of alkali–silica reaction gel by high-resolution NMR spectroscopy. *Journal of Non-Crystalline Solids*. v. 352, p. 3429-3436

3

Abrasion Resistance of High Performance Fabrics

Maja Somogyi Škoc and Emira Pezelj
Department of Materials, Fibres and Textile Testing
Faculty of Textile Technology, University of Zagreb
Croatia

1. Introduction

The second half of the 20th century was marked by the widespread use of textile materials in a variety of non-textile areas - from the application of fibre composites, high structural materials, a variety of materials and products used in industry, vehicles (cars, boats, planes, spaceships), professional sports, agriculture, road building and construction, protection of people and animals, environment protection, to medicine. Such textile materials are referred to as Technical Textiles, and are manufactured mostly of the fibres with some specific properties, such as high-tech fibres that under the environmental conditions retain their physical characteristics for a longer time, characterized by their high quality. These new generations of high-quality textile materials, manufactured employing most advanced technological processes, offered the road to making fabrics characterized by enhanced physical and mechanical properties and resistance to different impacts in use, while at the same time retaining their textile properties, such as colour fastness, dimensional stability, strength, good resistance to abrasion, etc. These materials, in the form of clothing, textile cartridges for shoes or any other end-use product, offer the retention of good looks, and thus extended usage. However, when it comes to satisfying the needs of soldiers, policemen, mountaineers and many other specific users, whose life depends on good abrasion resistance of textile materials they use, high performance acquires completely different meaning.

High performance materials are no longer a mystery; it is just a matter of time when the next generation of better and smarter fabrics will appear. As in everyday life, people like to be surrounded with textile materials that retain their original properties for a long time, together with high wear resistance under the conditions of use, i.e. high resistance to abrasion. It is necessary to choose the best method and appropriate procedures to test them, as well as the manners of expressing test results. This may be a good starting point for the development of new methods and test procedures, as well as an impulse for the construction of appropriate measurement equipment.

Today, there are several different types of apparatuses for testing abrasion resistance. They have evolved over time to include different kinds of loading conditions and materials, in order to be truer and closer to real conditions. Their results are not mutually comparable and often opposing results have been reported using different instruments. Lately, resistance to wear has been most common tested by Martindale, while the processes of circular wear with the tangential direction by Schopper have been mostly abandoned.

However, testing abrasion resistance by Schopper has its advantages and disadvantages. The use of SiC paper provides intensive wear at lower number of cycles and tangential contact sample with the abrading material. When using the Martindale method, the sample moves according to the Lissajous figure, standard wool fabric being abraded over the entire surface and this certainly contributes to getting reliable results.

This method is ideal for everyday fabrics, but when it comes to the fabrics in the high performance fabrics group, the method can be modified, which we discussed before [Somogyi et al. 2008].

Why is this important? It is well known that a textile material with high resistance to abrasion, such as Cordura®, can stand 100 thousand rubs by Martindale without a mass loss or specimen breakdown. Obviously, this kind of material will abrade a woven wool fabric, and not the other way around. Thus, the question arises whether the determination of Cordura® abrasion resistance, or some other the similar high performance fabric, using the Martindale methods, is suitable or the methods should be modified in order to obtain significant results?

In addition to the modification of Martindale in the manner described before [Somogyi et al. 2008], some modifications of standard methods of testing abrasion resistance by Martindale are also possible, all of them aiming at obtaining significant results. The Department of Materials, Fibres and Textile Testing, the Faculty of Textile Technology, University of Zagreb, has been involved in testing high quality fabrics intended for military and police use for some time. A number of high performance fabrics have been investigated and the need to determine wet abrasion resistance showed to be one of the key requirements throughout the investigation.

Fabrics intended for military and police uniforms are exposed to a number of physical and chemical agents in the course of regular use, rain and moisture being most frequently encountered. This means that, apart from testing dry abrasion resistance, as required by current standard, standard should be complemented by including testing wet abrasion.

Such a method for determining abrasion resistance is particularly suitable for damage detection, as damaged textile materials show more pronounced reduction in strength in wet state. The advantage of the modification is that the testing proposed can be done using the same methods, procedures and equipment as with dry materials. Similar to the above, permeability of air and resistance to water with good resistance to abrasion are very important for high performance fabrics used by mountain climbers, soldiers, policemen, firemen, etc. The Martindale method can again be ideally modified to suit the purpose; using the same apparatus, procedure and slightly modified methods it offers proper insight into the influence of wear on water or air permeability. Additional knowledge on the impact of testing abrasion resistance, as related to air permeability and water resistance, will be acquired in this way as well.

In general, the development of the materials today is not followed by the adequate development in testing methods and testing equipment. Numerous testing laboratories are in problems when testing high-performance fabrics. The authors wish to demonstrate the necessity to develop methods and equipment that would be able to follow the development of contemporary high-performance fabrics.

1.1 Abrasion resistance

For a long time it was considered that the tests of wear resistance i.e. abrasion resistance of textiles was a procedure providing an assessment of product durability i.e. its suitability for the intended purpose. However, over time, research has shown that it is not always so, but this does not mean the procedure has lost any of its importance. This is confirmed by the development of measuring techniques over the years, aimed to get the truer results, closer to reality. Eventually it has become clear that it is not even theoretically possible to construct a machine for measuring fabric wear generally, and systems should be improved for measuring wear of fibres, yarns and fabrics for specific loadings and for each type of textiles.

Abrasion occurs with textiles (test bodies) running relatively to some means of resistance, and is caused by friction, resulting in textile material wear. Investigations in real time can only deal with textiles and/or abrading agents, simultaneously or separately. The process of wear i.e. abrasion can proceed for some time with no visible damage. The term "to abrade" is used for this procedure. If the process is carried out to visible damage or failure of the material, the appearance of holes in the fabrics, then the expression "specimen breakdown" is used [Čunko R. 1989].

Resistance to abrasion is evaluated measuring the following values:

- loss of mass that occurs after a particular procedure of tear,
- the loss of material strength after the tear,
- the increase in air permeability after the tear in the fabric,
- the increase of light bandwidth after the tear in the fabric,
- the reduction of thickness in tear and
- the appearance of the worn surface (number of loops, thickening, lumps, etc.) [Čunko R. 1989].

The wear of textiles, abrasion, and hence the results of the tests, are affected by numerous factors related to the textile material, the environment in which the tests are conducted and testing conditions. Factors concerned with textile material are:

- fibre type,
- fibre properties,
- yarn twist,
- fabric structure and
- surface characteristics (hairiness, smooth, finishing, etc.).

The size of testing area is of key importance together with humidity and temperature, meaning that testing should be performed in the standard atmosphere for testing. The most important factors regarding the procedure of performing tests are:

- abrasion type,
- abradant type,
- pressure,
- speed,
- tension,
- the direction of abrasion,
- test tube tension,
- test tube carrier types,
- the duration of wear and
- the removal of the dust arising from textile materials [Čunko R. 1989; Saville B.P. 1999].

The impact factor of this group of factors is particularly high. It should be noted that the conditions of carrying out tests employing different methods differ significantly. It is therefore impossible to compare test results obtained by different methods, as well as the results obtained in any particular procedure, if the test conditions were not equal. Therefore, the results of testing relative wear are mostly descriptive, although they are of a significant general importance.

In practice, the most precisely tested aspect of fabric abrasion resistance, in assessing product performance in use, is its durability, since in most cases testing deals with clothing exposed to wear during use, or with fabrics designed for bed linen, furniture, technical textiles and others abraded under similar conditions. A range of test procedures and related equipment has been developed, mainly classified according to the method of performing the tests, the manner of staring and wearing the body of the sample, and the method of wear. In practice, most commonly used methods is circular blade wear, whether on a permanent contact throughout the test area, or in contact with certain parts of the surface during the procedure, with no preferential direction of the tear.

2. Experimental

2.1 Tested samples

High performance fabrics for different end-uses - military (sample no. 1, 2 and 3) and police (sample no. 4 and 5) were used.

Samples no. 1, 4 and 5 had a polyester membrane. It was a closed hydrophilic polyester membrane with pores that could not be penetrated by liquid water, but water molecules could be transported from the inside to outwards, employing adequate physical and chemical processes. The advantage of using such membranes is that they can be completely recycled. This type of sample is known as laminated textiles.

The sample no. 2 was classified as a super durable polyamide fabric. The sample no. 3 was made from super durable polyamide yarns. Some basic characteristic of the textile were defined according to the standards, as listed in Tab. 1, before modifying abrasion resistance according to the Martindale method.

Before sampling, test specimens were exposed to the standard atmosphere for 24 hours, in accordance with ISO 12947-3 and the modified method, free from tension, on a smooth horizontal surface. The standard temperate atmosphere for conditioning and testing textiles was used, e.g. the temperature of (20 ± 2) °C and the relative humidity of (65 ± 4) %, in accordance with ISO 139.

Testing parameters	Test method	Sample				
		1	2	3	4	5
Raw material content: -face -membrane -back	NN 41/2010 HRN ISO 1833: 2003	PES PES PES	PA	Cotton/PA	PES PES PES	PES PES PES
Mass per unit area [gm^{-2}]	HRN ISO 3801:2003	226.2	382.7	225.1	193.8	279.4
Thickness [mm]	HRN EN ISO 5084:2003	0.48	0.61	0.40	0.34	0.99
Breaking force [N] - warp - weft	HRN EN ISO 13934-1: 2008	1286 980	3537 3163	1119 829	1294 1181	816 619
Breaking elongation [%] - warp - weft	HRN EN ISO 13934-1: 2008	30.22 44.68	38.77 44.56	27.48 17.35	/ /	38.2 22.2
Purpose		army			police	

Table 1. Characterisation of the used samples

2.2 Abrasion resistance

Abrasion resistance of high performance fabrics by the Martindale method was done in accordance with the HRN EN ISO 12947-1: 2008, Textiles - Determination of the abrasion resistance of fabrics by the Martindale method, Part 1: Martindale abrasion testing apparatus, applicable for the testing of:

- woven and knitted fabrics,
- pile textiles having pile height of up to 2 mm and
- nonwovens.

The standard defines the requirements for the test apparatus according to Martindale and the auxiliary materials in order to determine the resistance of textile fabrics to wear. Martindale tester consists of the apparatus and accessories (Fig. 1). The equipment is easy to operate, but requires additional time to prepare the samples and perform other preparation tasks.

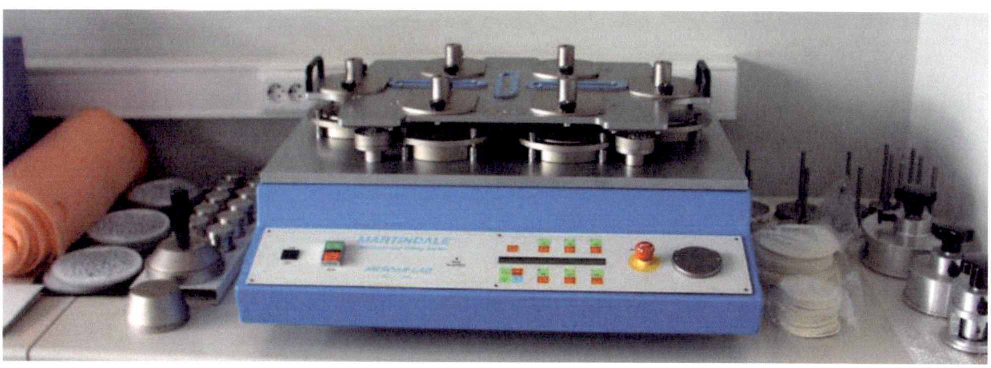

Fig. 1. Martindale abrasion testing apparatus

Martindale abrasion testing apparatus is designed to give a controlled amount of abrasion between fabric surfaces at comparatively low pressures in continuously changing directions.

ISO 12947 consist of the following parts under the general title - Textiles – Determination of the abrasion resistance of fabrics by the Martindale method:

- Part 1: Martindale abrasion testing apparatus,
- Part 2: Determination of specimen breakdown,
- Part 3: Determination of mass loss,
- Part 4: Assessment of appearance change [HRN EN ISO 12947-1:2008].

Abrasion resistance of high performance fabrics was tested according to the HRN EN ISO 12 947–3: Textiles – Determination of the abrasion resistance of fabrics by the Martindale method – Part 3: Determination of mass loss.

Specimen mass loss was determined for predetermined number of rubs. In our case, mass loss was determined at the following number of rubs: 1000, 5000, 25 000, 50 000 and 100 000. Mass loss was determined for each test, to the nearest 1 mg from the difference between test specimen mass before testing and the mass after testing.

Abrasion resistance of high performance fabrics by the Martindale method was first determined strictly in accordance with the ISO 12 947-3, and then the method was modified in two different ways:

1. determining abrasion resistance in wet state, and
2. determining abrasion resistance with separately prepared samples and wool abradant fabric for later determination of permeability (water and air).

Abrasion resistance of 5 different high performance fabrics was determined by the Martindale method. Modifications were carried out on the Martindale abrasion testing apparatus regarding different ways and conditions from the standard method. In both methods, mass loss (gravimetrically) and abraded surface appearance (microscopically) were determined. Microscopic examination was performed by reflected light using the universal microscope Dino-lite, with the magnification 60x, suitable for conventional laboratory control. Water and air permeability were provided according to the ISO standards too.

Modifications of the Martindale method were very important, especially the modification for the work in wet state, which was to serve as a basis for the determination of relative mass loss in wet state (f_w).

Relative mass loss in wet state could be determined after determining high performance fabric abrasion resistance in dry and wet state. We considered it highly important and calculated it using the equation (f_w):

$$f_w = \frac{m_{wet}}{m_{dry}} \cdot 100 \ [\%] \qquad (1)$$

As far as we knew, there has been no way devised to calculate relative wet mass loss (f_w) for so far. In this respect, the obtained knowledge could be a good and reliable basis for the improvements in abrasion resistance testing.

2.2.1 Abrasion resistance in accordance with the standard

Determination of abrasion resistance of high performance fabrics was done in accordance with the ISO 12947-3 as described above.

The Martindale abrasion tester uses a circular specimen, 38 mm in a diameter, cut by the appropriate cutter to a defined load, and rubs it against an abrasive medium (wool abradant fabric) in a translational movement, tracing a Lissajous figure, as shown in Fig. 2 [HRN EN ISO 12947-1:2008]. Lissajous figure is a motion resultant from two simple harmonic motions at right angles to one another.

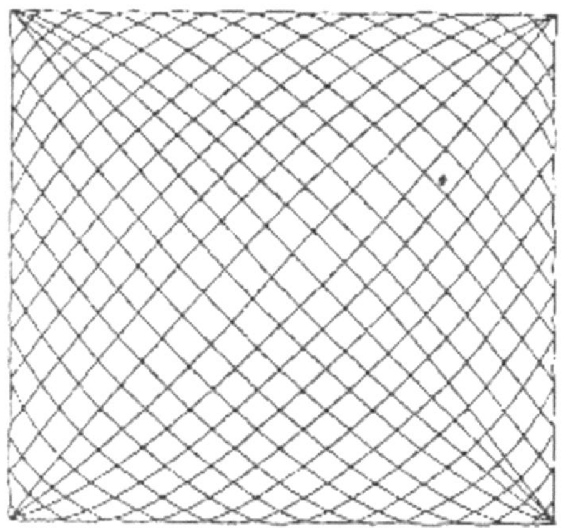

Fig. 2. Lissajous figure [HRN EN ISO 12947-1:2008]

The specimen was mounted in the holder with a circle of standard foam. The standard foam was a polyetherurethane foam material used as underlay for the test specimen, with the mass per unit area lower than 500 g/m^2. The foam was 38 mm in diameter, placed between the test specimen and the specimen holder insert.

Woven wool felt was placed on the abrading table and wool abradant fabric on the felt. They were pressed with a weight of the mass of (2.5 ± 0.5) kg and the diameter of (120 ± 10) mm and the ring was clamped with a clamping mechanism [HRN EN ISO 12947-3:2008].

Before determining abrasion resistance of high performance fabrics, Lissajous figure was checked for each work station by the method described in the annex A (ISO 12947-1).

The mass of the specimen holder and adequate loading piece was (795 ± 7) g, and the nominal pressure of 12 kPa (fabrics for technical use) was used, depending on the purpose of the high performance fabrics tested.

The specimens were subjected to abrasive wear for a predetermined number of rubs:

- 1000,
- 5000,
- 25 000,
- 50 000 and
- 100 000.

The mass of each conditioned test specimen was determined to the nearest 1 mg. Dry mass loss was determined according to the HRN EN ISO 12947-3, Textiles -- Determination of the abrasion resistance of fabrics by the Martindale method -- Part 3: Determination of mass loss.

2.2.2 Abrasion resistance in wet state

Abrasion resistance of high performance fabrics was determined in accordance with the ISO 12947-3, but not in all details. Before determining abrasion resistance of high performance fabrics, Lissajous figure was checked for each work station by the method described in the Annex A (ISO 12947-1), the same as when determining it in dry state (HRN EN ISO 12947-3:2008).

Specimens were first cut with the same cutter as the foam and then placed in Petri dish. Distilled water with a softening agent was in the Petri dish. The specimens spent 5 minutes there, prior to every 5000 rubs (Fig. 3). After that, the specimens were mounted in the specimen holders, with a circle of standard foam, as described in the standard (ISO 12947-3) and in 2.2.1.

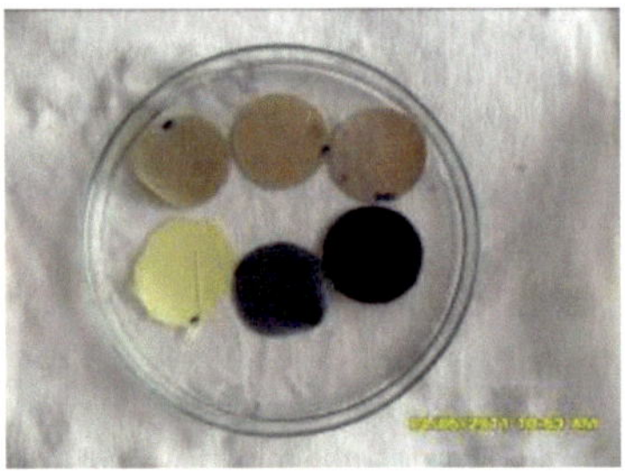

Fig. 3. Samples during wetting

Woven wool felts, wool abradant fabric and foam were also placed in the Petri dish, together with distilled water and a softening agent, prior to every 5000 rubs (Fig. 4).

Fig. 4. Auxiliary materials during wetting

After 5 minutes, woven wool felts, wool abradant fabric and foam were placed on cellulose wadding, in order to remove excess liquid.

Woven felts were then placed on the abrading table, protected with a transparent film. Wool abradant fabrics were placed on them and pressed with the weight of the mass described in 2.2.1. (Fig. 5).

Fig. 5. Abrading table with a transparent film before clamping the ring

The specimens were subjected to abrasive wear for a predetermined number of rubs, the same as in determining abrasion resistance, in accordance with the standard:

- 1000,
- 5000,
- 25 000,
- 50 000 and
- 100 000.

The mass of each conditioned test specimen was determined to the nearest 1 mg, before and after wetting. Mass loss in wet state was determined according to the HRN EN ISO 12947-3. Relative mass loss in wet state (f_w) was also determined.

Martindale abrasion tester for the determination of abrasion resistance in wet state was prepared as shown in Fig. 6.

Fig. 6. Martindale abrasion tester prepared for wet abrasion

2.2.3 Abrasion resistance and permeability

Abrasion resistance was determined using previously prepared samples and wool abradant fabric, for the purpose of determining air permeability and water resistance. The procedure partly followed the requirements of the ISO 12947-3. The chief reason for the modification of the Martindale method was again to obtain a sample with the diameter adequate to determine air permeability and water resistance.

The specimen was placed on a woven wool plate, instead of wool abradant fabrics. They were pressed with the weight of the mass of (2.5 ± 0.5) kg and diameter of (120 ± 10) mm in clamping the ring with a clamping mechanism, in the same manner as for dry abrasion resistance.

Wool abradant fabric was mounted in the specimen holder (instead of the specimen) with a circle of standard foam. The mass of the specimen holder and appropriate loading piece was (795 ± 7) g, with the nominal pressure of 12 kPa (fabrics for technical use), used depending on the purpose of the high performance fabrics tested.

The specimens were subjected to abrasive wear for a predetermined number of rubs (the same as for dry and wet abrasion testing):

- 1000,
- 5000,
- 25 000,
- 50 000 and
- 100 000.

The mass of each conditioned test specimen was determined to the nearest 1 mg. Dry mass loss for the sample (abrasive) and wool abradant fabric (as a sample) were determined according to the *HRN EN ISO 12947-3, Textiles -- Determination of the abrasion resistance of fabrics by the Martindale method -- Part 3: Determination of mass loss.*

2.3 Determining air permeability

The permeability of the samples to air, after abrasion, was determined in accordance with the *HRN EN ISO 9237:2003*, Textiles -- Determination of the permeability of fabrics to air (ISO 9237:1995; EN ISO 9237:1995).

The test samples were conditioned prior to testing in standard atmosphere for testing. Standard permeability here meant the rate of air flow passing perpendicularly through a given area of fabric. It was measured at a given pressure difference across the fabric test area over a given time period.

This study complied with the recommended test conditions, meaning test surface area was 20 cm^2 and pressure drop 200 Pa (for technical textiles). For some samples, for which we kept standard notes on pressure, an alternative pressure drop of 50 Pa or 500 Pa was used.

Arithmetic mean of the individual readings and the coefficient of variation were calculated. Air permeability, R, was expressed in millimetres per second, and was calculated using the following equation [HRN EN ISO 9237:2003]:

$$R= \frac{q_v}{A} \cdot 167 \qquad (2)$$

where
q_v is the arithmetic mean air flow rate (l/min),
A is the area of the fabric tested (cm^2) and
167 is the conversion factor from cubic decimetres (or litres) per minute per square centimetre to millimetres per second.

2.4 Determining water permeability

Water permeability was determined in accordance with the HRN EN 20811:2003, Textile fabrics -- Determination of resistance to water penetration -- Hydrostatic pressure test (ISO 811:1981; EN 20811:1992).

The hydrostatic pressure test method measured the resistance of the fabric to the penetration of water under hydrostatic pressure [HRN EN 20811:2003]. The test material was subjected to steadily increasing water pressure on the face, until water penetrated to the opposite face in three separate locations. These were identified as first, second and third water droplet. The pressure at which water penetrated the material was recorded in millibars, where the higher millibar value meant higher material resistance to water penetration. Millibars were transformed to Pascals for easier monitoring of the results.

3. Results and discussion

3.1 Results of abrasion resistance in accordance with the standard

After implementing the described method, the abraded samples were analyzed for mass loss and abraded surface, as shown in Tab. 2.

Table 2 shows that the samples no. 1 and 5 had a high loss of mass, in correlation with surface appearance. This is even more noticeable for the sample no. 5. At the beginning of testing (1000 rubs), the surface was uniform, with no protruding hairs of fibres, and at the end (100 000 rubs) the face of the fabric was smooth but not abraded as "conventional" fabrics; the result that could be seen is a membrane with smooth face.

The sample no. 1 was abraded too, but this was not as obvious from surface appearance as was from the results of mass loss. However, compared with the results of "conventional" fabric wear, this mass loss and the mass loss of the sample 5 were negligible (Tab. 4).

Sample no. 4 had visually distinct larger pore openings after 100 000 rubs and little mass loss, but both of these parameters were not significant.

The rest of the samples, especially the sample no. 2 and 3 did not exhibit any significant mass loss or change in surface appearance. Thus was expected, as the sample no. 2 was classified as a super durable polyamide fabric and no. 3 was made from durable polyamide yarns.

This determination of dry abrasion resistance in accordance with the standard showed that the samples such as no. 2, 3 and 4 exhibited little, practically negligible mass loss and change in surface appearance, so the method could not be considered suitable for the purpose. Why? Out experience indicates we could wear the samples for much more rubs (i.e. 500 000 or even more) and still surface appearance change and mass loss would be negligible. Such samples require a different test, either in terms of apparatus or the method itself. However, since the method according to Martindale is so widespread, there is no need to invent a new method, when this one could be easily modified. The additional advantage is retaining the advantage of wearing the fabric by the Lissajous figure motion, while obtaining results comparable to those obtained in actual use.

Sample /Testing parameters		1000 rubs	5000 rubs	25 000 rubs	50 000 rubs	100 000 rubs
1	Δm [%]	0.07	-0.18	-0.85	-4.61	-9.48
	Surface appearance	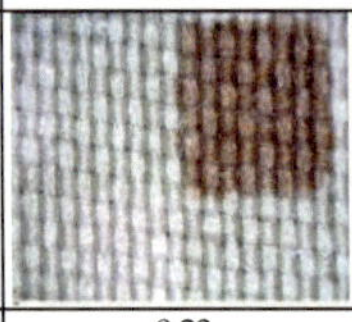				
2	Δm [%]	-0.23	0.36	1.09	1.20	1.04
	Surface appearance	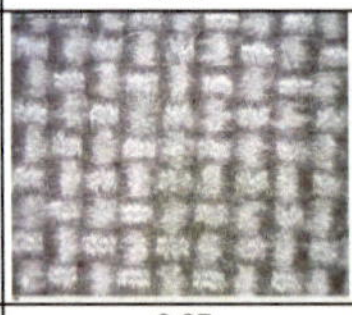	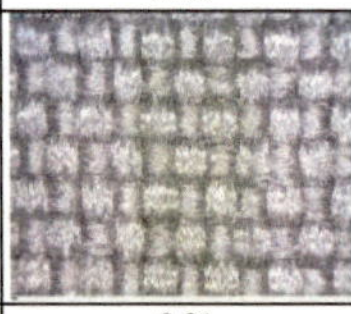	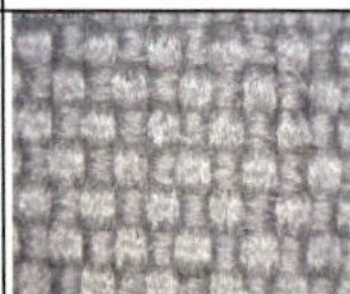		
3	Δm [%]	-0.85	-0.81	-0.95	-0.93	-2.08
	Surface appearance	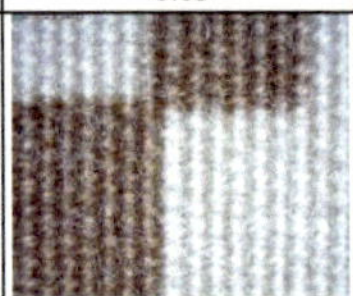	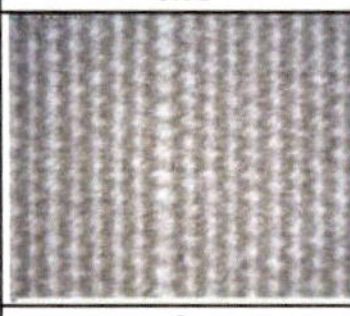	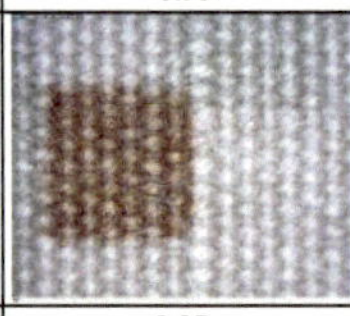		
4	Δm [%]	0.13	0	0.25	0.33	0.50
	Surface appearance	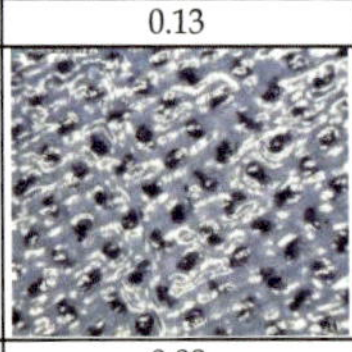	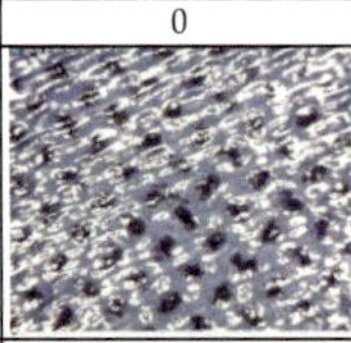	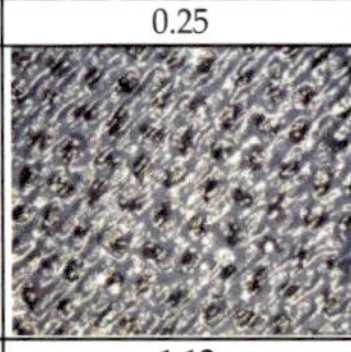	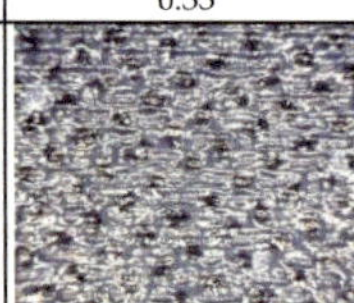	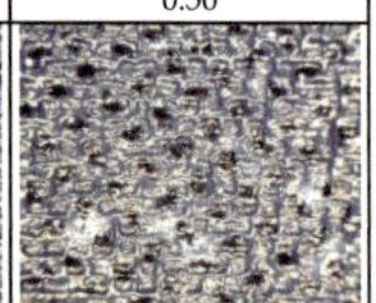
5	Δm [%]	0.09	0.29	-1.12	-3.80	-15.11
	Surface appearance	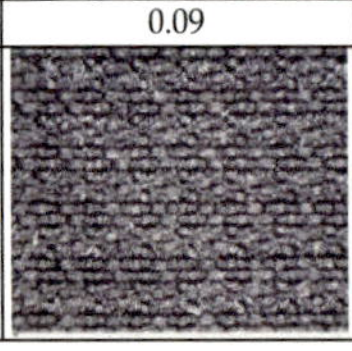	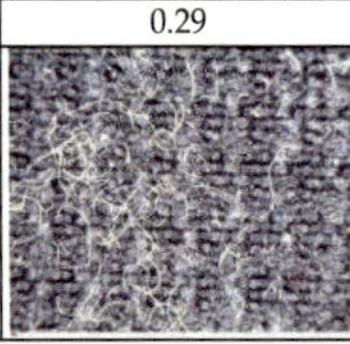	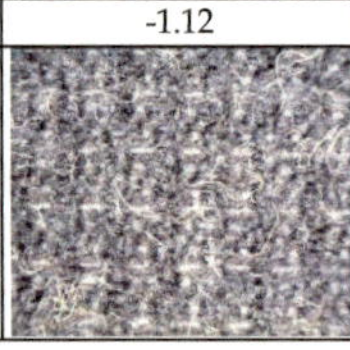	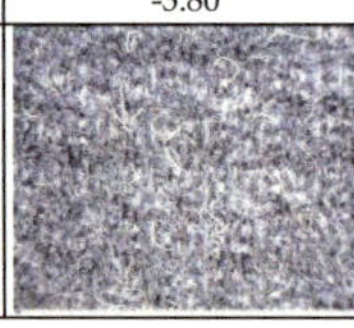	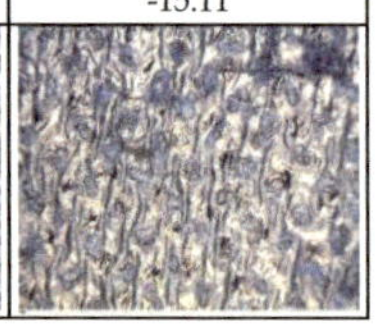

Δm minus (-) means mass loss, while plus (+) means mass increase.

Table 2. The results of mass loss after abrasion in dry state

3.2 Results of abrasion resistance in wet state

After employing the described method, the abraded samples were analyzed for mass loss and abraded surface as shown in Tab. 3.

Sample / Testing parameters		1000 rubs	5000 rubs	25 000 rubs	50 000 rubs	100 000 rubs
1	Δm [%]	-13.89	-8.57	-46.63	-16.34	-26.17
	Surface appearance	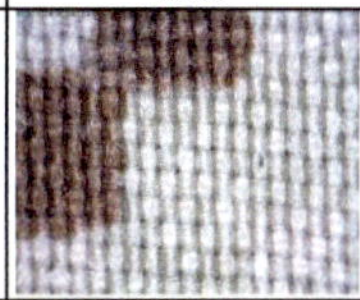				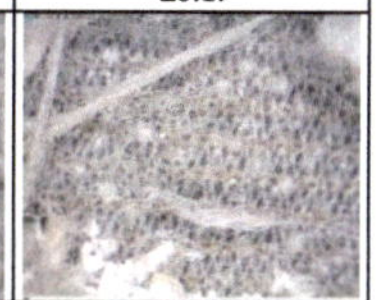
2	Δm [%]	-1.02	6.43	-21.76	-2.67	-26.49
	Surface appearance	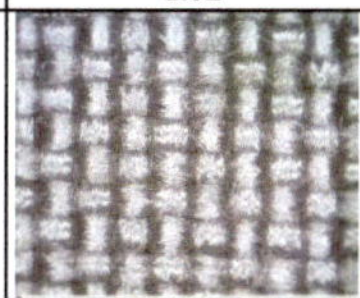	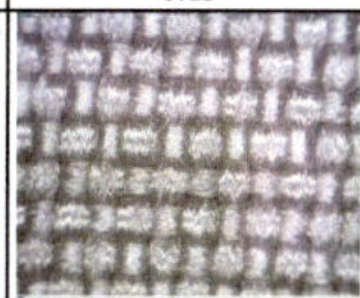	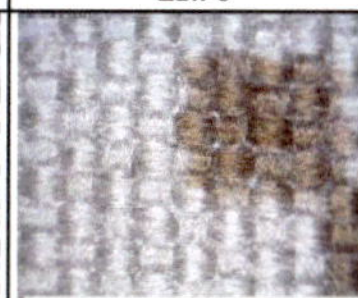	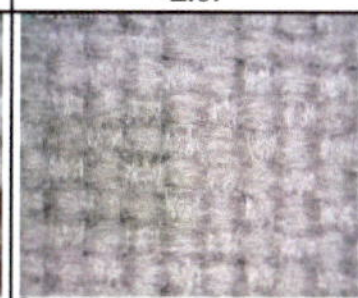	
3	Δm [%]	-7.66	-0.18	-45.19	-36.54	-70.25
	Surface appearance	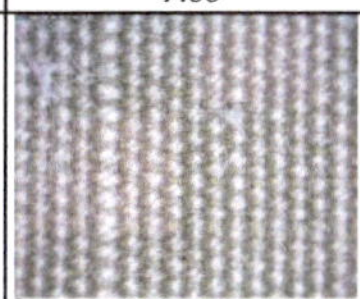			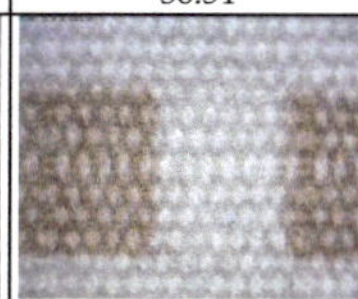	
4	Δm [%]	-16.63	-21.79	-34.09	-5.46	-38.80
	Surface appearance		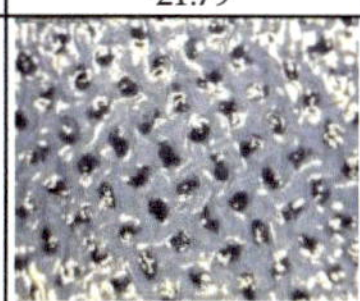	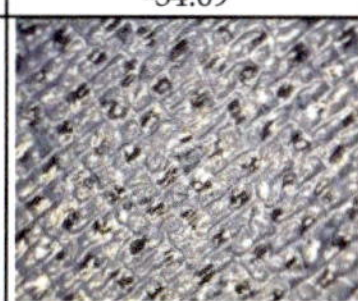	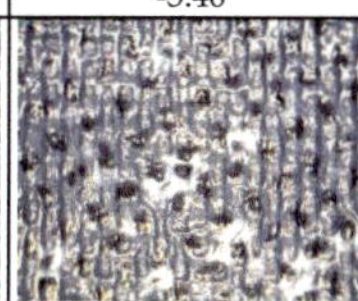	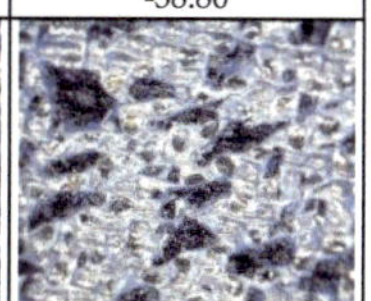
5	Δm [%]	-39.61	-48.39	-80.62	-43.42	-86.12
	Surface appearance		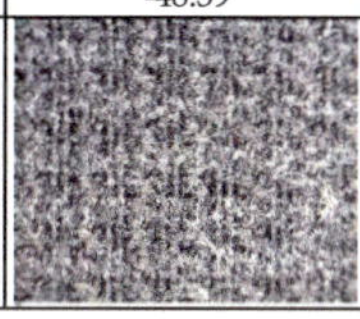	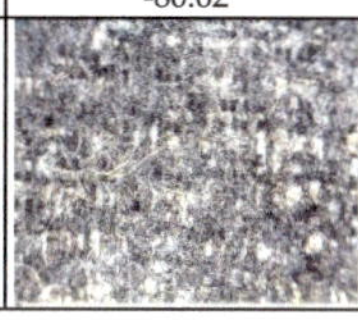	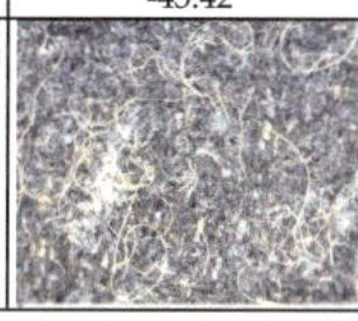	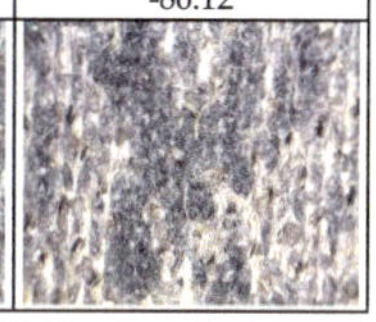

Δm minus (-) means mass loss and plus (+) means mass increase

Table 3. The results of mass loss after abrasion in wet state

Table 3 showed that testing with the modified Martindale method in wet state was more rigorous. All the tested samples exhibited significant mass loss in the range from 1000 to 100 000 rubs.

Such behaviour was exhibited even by the sample no. 2, which was classified as a super durable polyamide fabric and no. 3, made from durable polyamide yarns. Small remains of a wool abradant fabric were found on the surface of the sample no. 2, i.e. between the warp and weft threads. The sample no. 3 had only warp threads after 100 000 rubs, while weft threads were broken.

The sample no. 1 had fully abraded face and the membrane could be seen at 100 000 rubs.

Pores did not exist anymore on the face of the sample no. 4, after 100 000 rubs. They were smoothened out to the membrane and the back of the sample.

Surface appearance of the sample no. 5, after 100 000 rubs, showed abraded face of the fabric as well as abraded membrane, so the back of the fabric could be seen.

Based on the results obtained for mass loss and surface appearance, it could be concluded that intensive wear occurred in wet state. This also showed that, when compared with dry testing, this method was much more significant, and therefore more suitable for testing fabrics intended for military and police use, particularly as the simulated conditions of wear at work were most often in wet state.

Testing performed clearly pointed to the need for testing high performance fabrics in wet state, since the changes in dry state were negligible. This is especially true when the fabrics were designed to protect human body under conditions other than ideal i.e. in everyday use.

The investigations of this type should be conducted for high performance fabrics only, because most "conventional" fabrics cannot withstand more than 5000 cycles of abrasion in wet state, as shown in Tab. 4.

Testing parameters /Sample	"Conventional" fabric for protective clothing	
Rubs	1000	5000
Δm [%]	-2.77	-3.70
Dry surface appearance		
Δm [%]	-61.83	-37.94
Wet surface appearance		

Table 4. The results of mass loss on "conventional" fabrics after abrasion in wet state

Relative mass loss in wet state was calculated in accordance with the equation no. 1. This type of test could be particularly interesting on the materials made from natural cellulose. Generally, they exhibit a greater loss of mass in the wet than in the dry samples. However, this investigation was concerned with synthetic fibres, and the results shown in Table 5 deal with them.

Sample	f_w [%]				
	1000 rubs	5000 rubs	25 000 rubs	50 000 rubs	100 000 rubs
1	24.9	13.2	81.1	18.2	0.5
2	1.2	7.4	29.5	2.7	36.8
3	14.9	35.5	102.8	55.6	168.3
4	20.4	31.5	54.4	5.8	63.5
5	164.0	251.2	453.5	72.3	511.9

Table 5. Relative mass loss in wet state

The sample no. 3 was a mixture of natural cellulose and polyamide fibres and exhibited a significant mass loss in wet state. However, comparing these results with the sample no. 5, we could see that the polyester sample had a higher mass loss.

Even the sample no. 4 exhibited a relatively high mass loss, as did the samples no. 3 and 5, while the mass loss for the sample no. 2 was slightly smaller.

The measurements on the sample no. 2 showed that super durable polyamide fabrics did not show superior wear resistance in the wet state, as it did in the dry state (Tab. 6).

These results confirmed the thesis that higher wear occurred in the wet than in dry state, for all the samples, regardless of the raw material composition. This is of high importance, proving this method could be generally applied.

Anyhow, the potential of the modified method could be further expanded and investigated, before considering the update of any other method or changing the standard.

Sample /rubs	Wet surface appearance				
	1	2	3	4	5
100 000 rubs	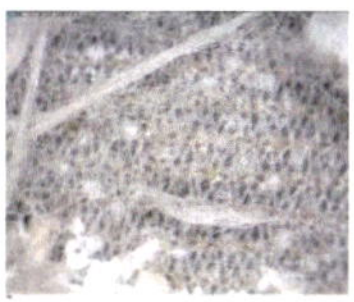			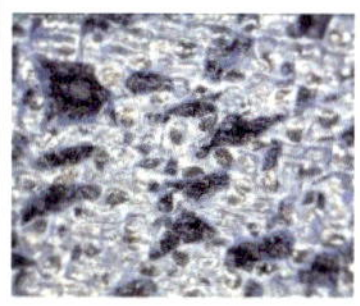	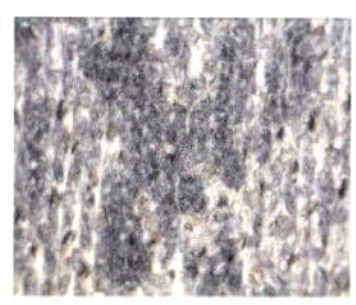
	Dry surface appearance				
	1	2	3	4	5
	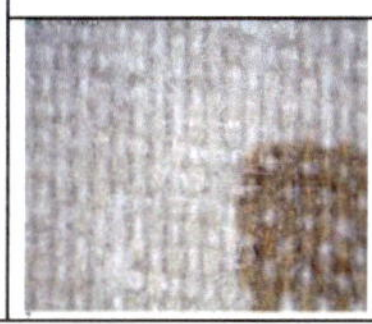	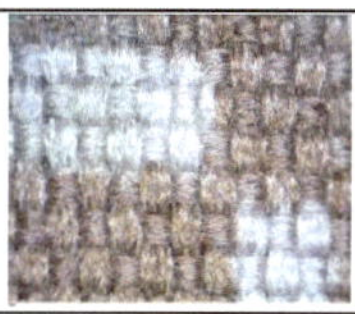	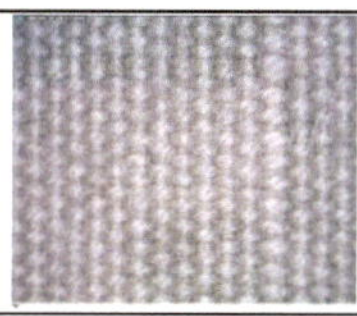	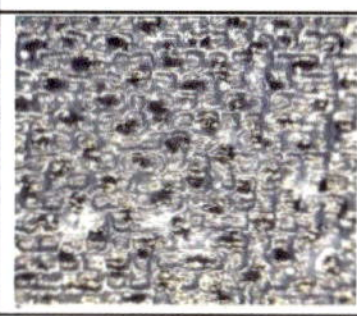	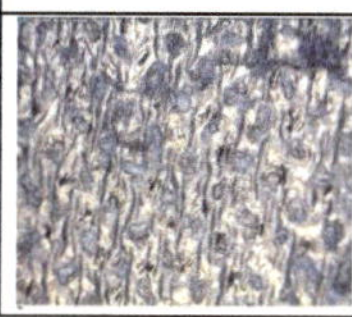

Table 6. Final comparison of surface appearance for wet and dry abraded high performance samples at 100 000 rubs

3.3 The results for the impact of abrasion resistance on air permeability and water resistance

After applying the prescribed methods, the abraded samples were analyzed to determine air permeability and water resistance. The results are presented in Tab. 7.

According to the results presented in Tab. 7 it can be concluded that this method is not as rigorous as the previous one.

The investigated samples of high performance fabrics retained their original properties. This suggested insufficient modification of the testing method employed. Specifically, wool abradant fabric was insufficient, i.e. too small (in the sample holder) to produce any significant result. Surface appearance of the samples in Tab. 7 confirmed this.

Sample / Testing parameters		1000 rubs	5000 rubs	25 000 rubs	50 000 rubs	100 000 rubs
1	Air permeability [mm/s]	3.2	3.8	4.3	3.5	2.4
	Water penetration [Pa]	> 100 000	> 100 000	> 100 000	> 100 000	> 100 000
	Surface appearance	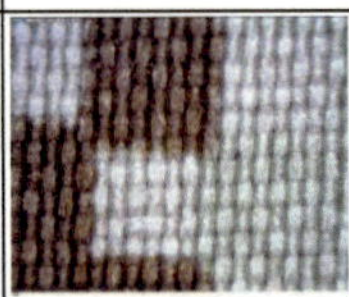	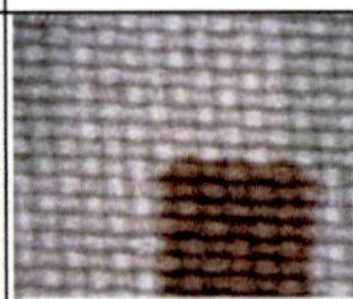	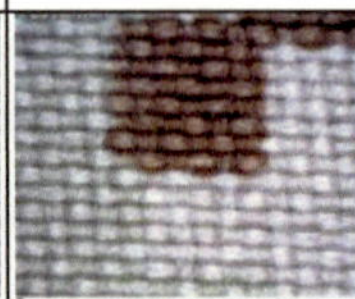	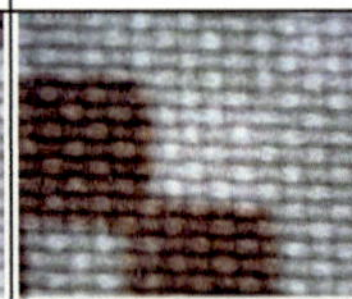	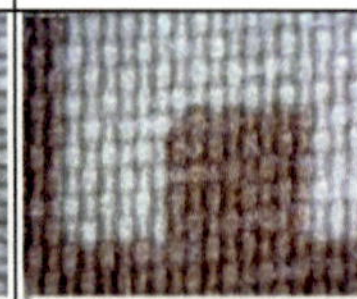
2	Air permeability [mm/s]	10.1	9.8	8.4	8.7	8.6
	Water penetration [Pa]	5 000	5 000	10 000	20 000	20 000
	Surface appearance		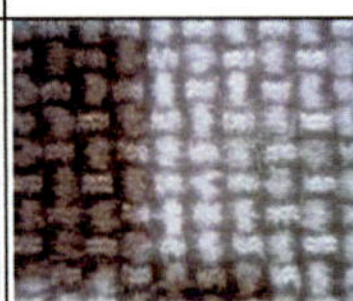	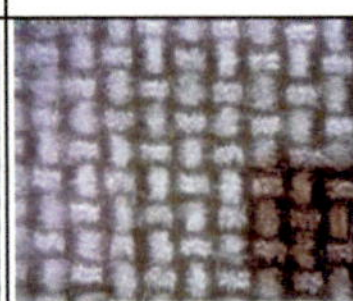	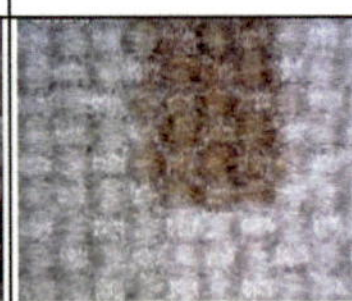	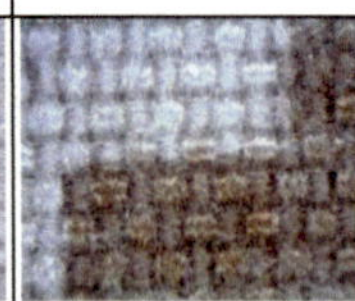
3	Air permeability [mm/s]	27.6	25.7	25.7	25.7	24.5
	Water penetration [Pa]	0	0	0	0	0
	Surface appearance			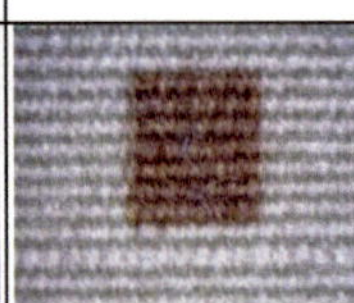		
4	Air permeability [mm/s]	impermeable	impermeable	impermeable	impermeable	impermeable
	Water penetration [Pa]	20 000	2 500	5 000	5 000	5 000
	Surface appearance			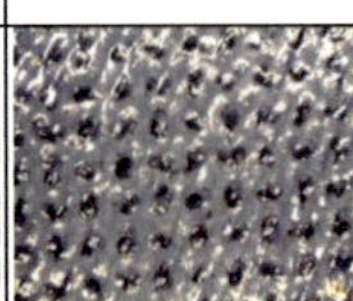		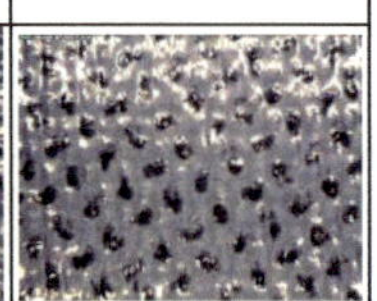
5	Air permeability [mm/s]	impermeable	impermeable	impermeable	impermeable	impermeable
	Water penetration [Pa]	> 100 000	> 100 000	> 100 000	> 100 000	> 100 000
	Surface appearance					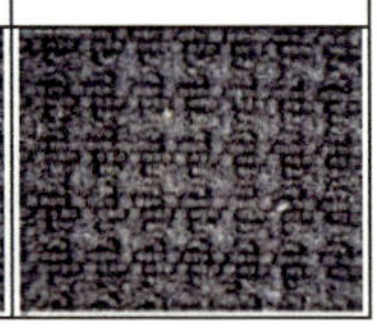

Table 7. The results of air permeability and water resistance after wear test

To achieve significant results, sandpaper or some other agent more active on the surface of the test sample should be used instead of wool fabric. The samples tested showed excellent water resistance and good air permeability characteristics. These results were confirmed by the appearance of the surface. Visual inspection using Dino-lite system showed no difference between the samples at 1000 rubs and the samples at 100 000 rubs.

4. Conclusion

The modifications of the standard Martindale method used to determine fabric abrasion resistance may result in knowledge of high performance material properties. The investigations presented offer scientific and practical contributions to the modification of abrasion resistance testing methods for high performance fabrics.

Original modifications have been carried out in this way and textile technology is offered a new and uncharted territory. Analysing mass loss and appearance of the abraded fabric surface it can be concluded that the standardized method for abrasion characterisation by Martindale is not suitable for high performance fabrics.

The results obtained indicate a more appropriate way of testing the resistance of high performance fabrics to abrasion, using the modified Martindale method in wet state, as presented in this chapter.

The tests were conducted on five commercially available textile products. These tests will contribute to the development of new methods of testing abrasion resistance of high performance fabrics.

As far as the authors know, the relative mass loss in wet state (f_W) has not been calculated so far. The knowledge obtained by the investigation presented can be a good and reliable basis for improving abrasion resistance tests in wet state.

The method according to Martindale is in widespread use and this method (or methods) can be easily modified for the above purpose. In this way, the advantage of wearing by the Lissajous figure will be retained, while, on the other hand, the results obtained will be quite near to those obtained in actual use of the fabrics.

5. Acknowledgment

Authors would like to thank the Ministry of Science, Education and Sport of the Republic of Croatia for the financial support of this study, which is a part of the research project „Multifunctional Human Protective Textile Materials"(117-1171419-1393).

6. References

Čunko, R. (1989). *Ispitivanje tekstila,* University of Zagreb Faculty of Textile Technology, ISBN 86-329-0180-X, Zagreb, Croatia

HRN EN 20811:2003, Textile fabrics -- Determination of resistance to water penetration -- Hydrostatic pressure test (ISO 811:1981; EN 20811:1992

HRN EN ISO 12947-1:2008; Textiles – Determination of the abrasion resistance of fabrics by the Martindale method, Part 1: Martindale abrasion testing apparatus

HRN EN ISO 12947-3:2008, Textiles -- Determination of the abrasion resistance of fabrics by the Martindale method -- Part 3: Determination of mass loss

HRN EN ISO 9237:2003, Textiles -- Determination of the permeability of fabrics to air (ISO 9237:1995; EN ISO 9237:1995)

Saville, B.P. (1999). *Physical testing of textiles,* Woodhead Publishing Limited, ISBN 1 85573 367 6, Cambridge, England

Somogyi, M.; Pezelj, E.; Čunko, R.; Vujasinović, E. (2008). High performance Cordura®, fabrics - resistance to abrasion, *ITC&DC book of Proceedings of the 4th International Textile, Clothing & Design Conference, Magic World of Textiles,* pp. 882-887, Zagreb, Croatia, October 05th to 08th 2008

4

Numerical Simulation of Abrasion of Particles

Manoj Khanal[1] and Rob Morrison[2]
[1]*Queensland Centre for Advanced Technologies, Earth Science and Resource Engineering, Commonwealth Scientific and Industrial Research Organization, Technology Court, Pullenvale,*
[2]*Julius Kruttschnitt Mineral Research Centre, University of Queensland, Indooroopilly, Australia*

1. Introduction

Abrasion is a surface breakage event where irregular surfaces of particles are removed. It produces smoothed surfaces/particles by comminuting irregular edges. Depending on the application, abrasion can be a desired or undesired process, and may happen either naturally or intentionally. For example, in some industries where spherical particles are end products, abrasion can be a desired process but in some process industries it is undesired because it produces dust, which makes it difficult for handling, as well as increasing the potential for losses of material. In some cases, natural phenomena (weathering, time dependent effect, age, and contact) can abrade materials and sometimes abrasion of materials is enhanced by applying energy. The loss of surface layers of particles due to weathering can be considered as a natural abrasion process whereas the use of mills to generate fines are considered as abrasion via applied energy.

Continuous abrasion of materials also implies particle size reduction. When a particle is stressed, the particle response depends on its size and shape, and the conditions under which stress is applied. In general, the governing parameters for abrasion can be differentiated into three groups – material parameters, process parameters and machine parameters. The initial stress condition (history), diameter (scale), shape and size, homogeneity and type of materials are known as material parameters. The process parameters include type of stress, particle-particle configuration, stressing velocity and frequency, temperature, specific energy. Similarly, the machine parameters cover shape, size and hardness of the stressing tool, and size and geometry of the stressing machine. The requirement (shape and size) of the final abraded products depends on the better selection of these parameters.

The diversity and the purpose of tasks to be performed during abrasion have led to the development of a range of abrasion testing machines. There is no uniform criterion available for classifying abrasion and the system used for naming machines is not standardized. Hence, a particular type of machine may have several names. Similarly there is no any standard definition of mechanisms of size reduction ratio. Mostly it is considered as a slow process where a "significantly less" amount of materials are removed from the parent material. There is no definition of "significantly less" which classifies the process as abrasion. For example, if a parent particle loses 10% of its mass during fracture process then it is termed as comminution, if the outputs are lumps larger than a few centimetres then the process is defined as crushing and if the outputs are in finer and in millimetre ranges the process is called milling (Rumpf, 1995). A similar definition of abrasion is not available in the literature. In short the definition of abrasion is more of a concept rather than a mathematical definition.

There are various methods and machines, however in mineral processing tumbling mills are typically used, to study the abrasion. This paper is also based on the tumbling mill. A tumbling mill can be operated in a controlled environment, where basically, three different stages of particle failure can be observed. The first stage is surface failure where particles lose their edges and surfaces to become more rounded shapes. Surface breakage mechanisms like "abrasion" and "attrition" can be classified within this stage.

Surface breakage is followed by the second stage which involves body breakage of particles. In this stage, parent particles lose a large fraction of their mass and generate progeny. This can be termed as body breakage or, most commonly, comminution. The third stage can be termed as secondary breakage, where already comminuted particles disintegrate further. The fragments generated in the second stage further generate sub-fragments. These three stages depend upon the tumbling time and the intensity of stresses generated in the particles.

In milling, particles of interest are processed to produce surface breakage events. Secondary media like steel balls and/or pebble ports maybe used to modify the product size distribution. As the mill rotates for abrasion milling, primary media (rock particles), and secondary media (grinding balls) if present, tumble against each other and the material breaks mainly by abrasion. Extensive experimental work has been done by Loveday and Naidoo (1997), Loveday et al. (2006), Loveday and Dong (2000), Loveday and Whiten (2002) to study the abrasion of mineral particles. Most of these studies are empirical in nature. A few studies (Morrison and Cleary, 2004, Khanal and Morrison, 2008), have used the discrete element method (DEM) to study the abrasion behaviour of particles. There is no globally accepted model which describes the abrasion of particles in general.

The objective of this paper is to study the relevance of 3D Discrete Element Modelling (DEM) simulation of abrasion in small scale tumbling mills. The paper also focuses on DEM simulation to investigate ore abrasion resistance and evaluation of abrasion process parameters of spherical and non-spherical particles in a milling environment. Different mill sizes are selected for this study. Various process parameters, for example, effect of particle size, milling time and stiffness of particles on abrasion characteristics have been evaluated. The paper also discusses effect of surface roughness on the particle trajectories, collision force, collision energy, and energy spectra. These DEM results are compared with experimental results where applicable.

As the paper deals with the abrasion of particles, we wish to avoid body breakage. The grinding time for the abrasion process has been noted from the experiments (Banini, 2000) and it represents an abrasion only case, hence the descriptions are based on it. The DEM simulations have been carried out in order to support and analyse experimental interpretation with numerical simulations.

2. Experimental work

Experimental data were extracted from Banini (2000). Four different mill sizes of 0.2 m, 0.3 m, 0.6 m and 1.1 m internal diameter were manufactured. Each mill has rectangular lifters of 30 mm in height placed at regular intervals. The number of lifters was different in different mills. Table 1 shows the number of lifters present and the rotational speed of each mill. The rotational speeds for 1.1 m and 0.6 m diameter mills were 60% and for 0.3 m and 0.2 m mills were 70% of the critical speed. During each revolution, particles are lifted to a maximum point and then released under gravity. However the particles may reach their highest point after leaving contact with the lifter.

Mill Diameter	**1.1 m**	**0.6 m**	**0.3 m**	**0.2 m**
Mill Speed (rad/sec)	2.53	3.42	5.66	6.93
Number of Lifters	12	8	6	4
Size Fraction (mm)	**Release Heights (m)**			
90.0 - 75.0	0.94	0.46	0.24	0.13
63.0 - 53.0	0.97	0.52	0.24	0.14
45.0 - 37.5	0.97	0.53	0.27	0.14
31.5 - 26.5	0.97	0.54	0.27	0.16

Table 1. Size fraction, mill speed, mill diameter, release height and number of lifters used in experiments (Banini, 2000)

There are centrifugal and frictional forces when the particles are in contact with the liners and lifters. The falling of particles from the highest point will transfer maximum amount of energy to the particles during each revolution with a minimum level of projection/ trajectory (Banini, 2000). Table 1 also shows the maximum release heights for different mills and size fractions. For the largest particles (considered here) in the 1.1 m mill, 0.94 is the release height observed from the experiments, which is 85% of the height of the mill. Similarly, for 0.6 m, 0.3 m and 0.2 m mills, the release heights are 76%, 80% and 65%, respectively of the mill diameter. This increases with the decrease in size of the particles in the mill. This is due to the fact that the center of mass of particles lie closer to the mill shell and this provides relatively more stability to the smaller particles, hence they are lifted higher than the larger particles.

Several rock types were tested. The details of the rock types can be found in (Banini, 2000). Table 2 shows the number of particles tested per size fraction in each mill. Extended tumbling generates body breakage and produces large fragments. Particles which have been repeatedly impacted during the tumbling process generate incremental (progressive) damage, and subsequently break (body breakage) at low stressing energies. This was confirmed by Banini (2000), as tumbling time increased beyond 10 min the original particles began to undergo body breakage for the considered cases. Further, Loveday and Naidoo

(1997) have mentioned that particle size reduction results from grains loosened by multiple impacts rather than from rubbing of surfaces.

Size Fraction (mm)	Number of Particles
90.0 - 75.0	8
63.0 - 53.0	12
45.0 - 37.5	20
31.5 - 26.5	36

Table 2. Number of particles tested per size fraction in each mill (Banini, 2000)

Increasing the mill load leads to an increase in wear rate of particles (Loveday and Naidoo, 1997). It has been observed that the rock-rock collision and overcrowding at the toe of the mill results in production of unusually fine material (Banini, 2000). The rate of abrasion of rocks can be reduced by the presence of fines and/or chips (Loveday and Dong, 2000). Table 3 shows the number of rocks per charge for each mill and size fraction, respectively to reduce the rock-rock collision.

Size Fraction (mm)	Mill diameter (m)			
	1.1	0.6	0.3	0.2
90.0 - 75.0	4	4	2	2
63.0 - 53.0	6	6	3	2
45.0 - 37.5	10	10	4	2
31.5 - 26.5	18	18	6	4

Table 3. Maximum number of particles in each mill and size fraction to abrasion only case for rock-rock collisions (Banini, 2000)

3. Numerical simulations

In DEM, particles are considered to be distinct elements. The laws of motion and material constitutive laws are applied to each element. In this paper, DEM simulations were carried out with a commercially available package (EDEM, 2006). The Hertz-Mindlin (no slip) contact law has been used to model the contact interaction among entities. The milling condition and parameters used in the experimental work were simulated. The mills and lifters were designed in SolidWorks (2006) and imported into EDEM. The required number of particles for each simulation was generated by varying size and distribution of the sample particle. Tables 4a and 4b show the material properties assigned to particles and mills, and number and size of particles used for the simulations in different mills, respectively.

	Poissions ratio	Shear Modulus (GPa)	Density (kg/m^3)
Particle	0.25	0.1	2650
Mill	0.29	80	7861

Table 4a. Particle and mill properties for simulations

Loveday and Naidoo (1997) found that the abrasion rate per unit mass increases significantly with rock mass. In abrasion modeling it is difficult to relate experimental work to DEM simulation results. Therefore, different hypotheses and assumptions have to be made to the data available from the simulations to compare with experimental results. Generally, it was assumed that the number of collisions in the simulations is responsible for particle disintegration (abrasion) and each collision accounts for an incremental damage and certain amount of mass is comminuted from the parent specimen. In other words, the number of collisions (with suitable assumptions and relations) can be compared with experimental mass losses from the particles. A good correlation has been found between measured particle mass loss and the DEM estimate of cumulative frictional energy absorbed by the particles (Cleary and Morrison, 2004).

Simulation	**Mill size (m)**	**Number of particles**	**Particle size distribution (random distribution, mm)**
1	1.1	8	72 - 90
2		12	53 - 63
3		20	37.5 - 45
4		36	26.5 - 31.5
5		38	26.5 - 90
6		76	26.5 - 90
7	0.6	4	75 - 90
8		6	75 - 90
9		8	75 - 90
10		10	37.5 - 45
11		12	53 - 63
12		18	26.5 - 31.5
13		20	37.5 - 45
14		36	26.5 - 31.5
15		38	26.5 - 90
16	0.3	8	26.5 - 90
17		15	72 - 90
18	0.2	10	26.5 - 90

Table 4b. Numbers and sizes of particles used for simulations in different mills

The paper also deals with non-spherical particles. They were generated by combining different individual particles. The figure shown in Table 5 illustrates the complex particle which can be formed by combining many smaller particles. Rough surfaces of the template show the asperities and roughness present on the particle. The required number of particles for each simulation was generated by varying size and distribution of the template particle. Table 5 shows the assigned particle and mill material properties.

Different particle structures have been simulated to study the effect of surface roughness in abrasion. Eight different particles with random distributions in the range of 70 to 90mm have been simulated as the load in a 1.1m mill. The roughness of particles was varied from particles having 15 "humps" (asperities) to perfectly spherical particles without "hump." The template particle shown in Table 5 represents a particle structure with 15 humps. Table 5 shows the number of surfaces created on the template particle to simulate eight-particle systems in a 1.1m mill.

Rough surface	**Mass, kg**	**Remarks**	**Template of a particle**
Single sphere	7.57	planet	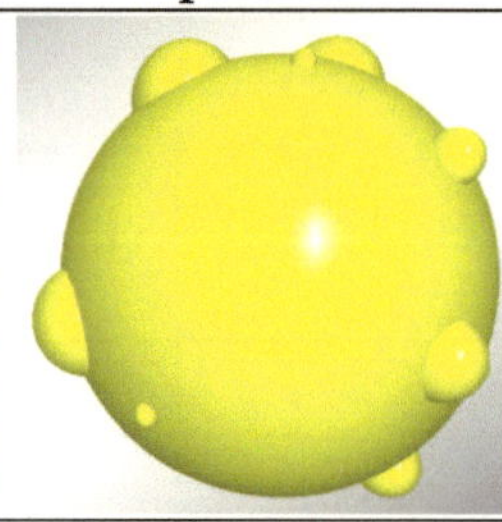
1	7.36	1_surface	
7	4.66	7_surface	
13	4.55	13_surface	
15	4.55	15_surface	
Particle and mill properties for simulations			
	Poisson's ratio	*Shear Modulus (GPa)*	*Density (kg/m³)*
Particle	0.25	0.1	2650
Mill	0.29	80	7861

Table 5. Particles having different types of rough surfaces used to simulate an 8-particle system in 1.1m mill

4. Results and discussion

4.1 Spherical particles

4.1.1 Particle trajectories

Figure 1 shows the results of DEM simulations of particle trajectories in the 1.1 m and 0.6 m mills for mills containing 8, 12, 20 and 36 particles. The figures suggest that the particles follow the circumference of the mill up to the maximum height the lifter can take. Then due to gravity, the particles fall within the mill. At the steady state condition, the lifters throw the particles as far as possible in a projectile trajectory. Depending on the population (density) of particles inside the mill, the fallen particles hit either another particle or the mill shell. For the lightly loaded conditions, the particles do not collide with each other before hitting the mill shell. But if the mill is overcrowded with particles, many inter-particle collisions will occur and finer particles will be generated. Hence, particle-particle interaction does not appear to occur until a critical number of particles and increase in mill loads increase in wear rate of the particles (Loveday and Naidoo, 1997; Banini, 2000).

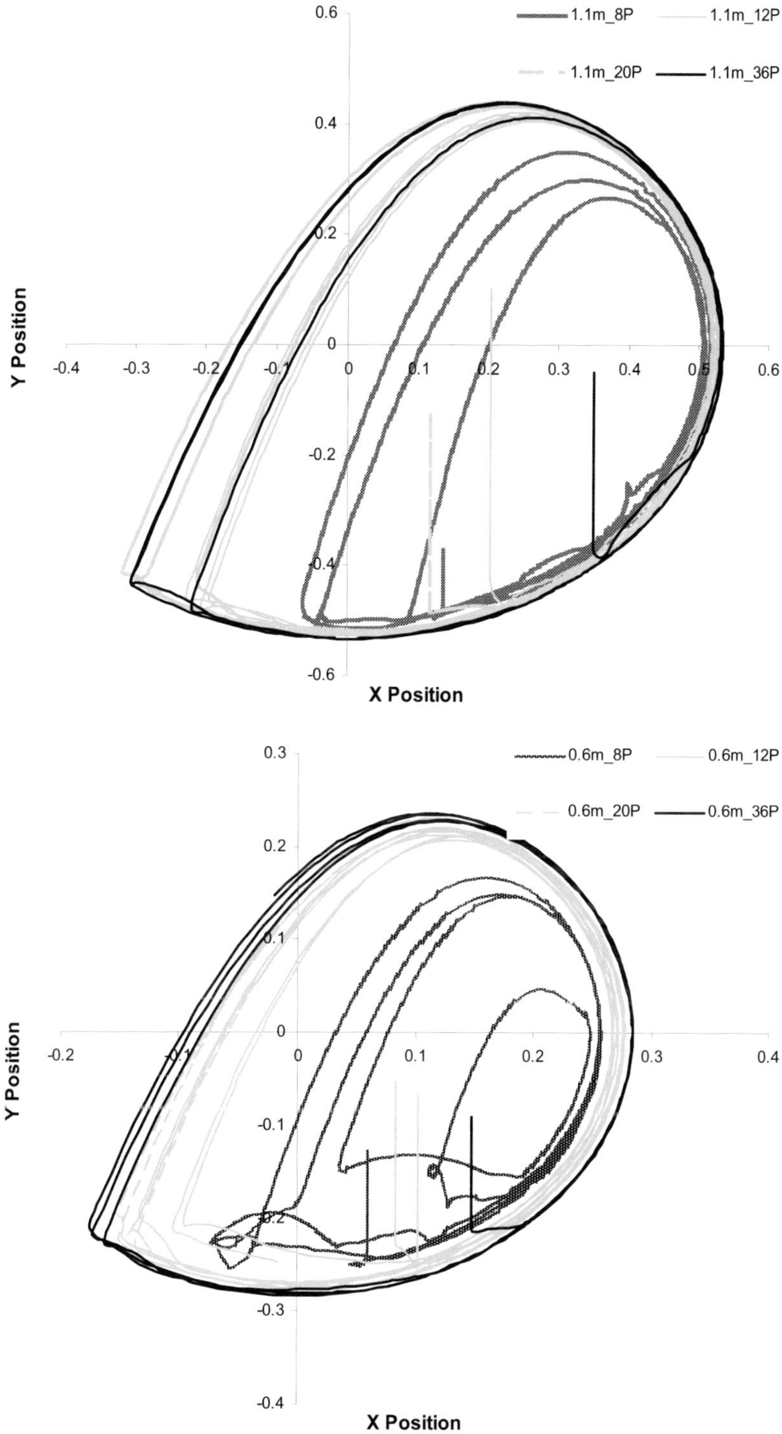

Fig. 1. Particle trajectories for 1.1 m (left) and 0.6 m (right) mills

4.1.2 Release height analysis

Particles on the lifters are lifted up to a certain height and release due to gravity. The lift-up and release heights are important as they impart potential and kinetic energy to the particles, as a result they govern the collision energy. It was noted by Djordjevic et al. (2006) that smaller particles are lifted higher than larger particles as their centres of mass are closer to the mill shell than the outer edge of the lifter bar.

The release heights were measured experimentally (Banini, 2000) by observing the motion of the charge through the perspex front cover of the mill. In the simulations, DEM allows the user to calculate the release height by tracking the simulated trajectories of the rock particle. The experimental release heights are shown in Table 1 for different mill sizes and size fractions. The maximum achievable heights of the particles in different mills during simulations are shown in Figure 2. The figure shows the height of the particles versus time for 1.1, 0.6, 0.3 and 0.2 m mills with 8, 10, 12, 15 and 36 particles. The average release height for 1.1 m mill with 36 particles is 0.94 m in simulations whereas 0.97 m is in experiments. Similarly, for 0.6 m mills, the average release height is 0.52 m in simulations and 0.54 m in experiments. The release height was calculated from the difference in position of the particles (along the height of the mill) in one revolution of the mill. The slight difference in the release height is due to the selection of particles to trace the height in experiments and simulations. The release height maybe different depending on the size of the particles selected to calculate the release height as size and shape (for the position on lifter) of the particles influence the lift-up height (Djordjevic et al., 2006).

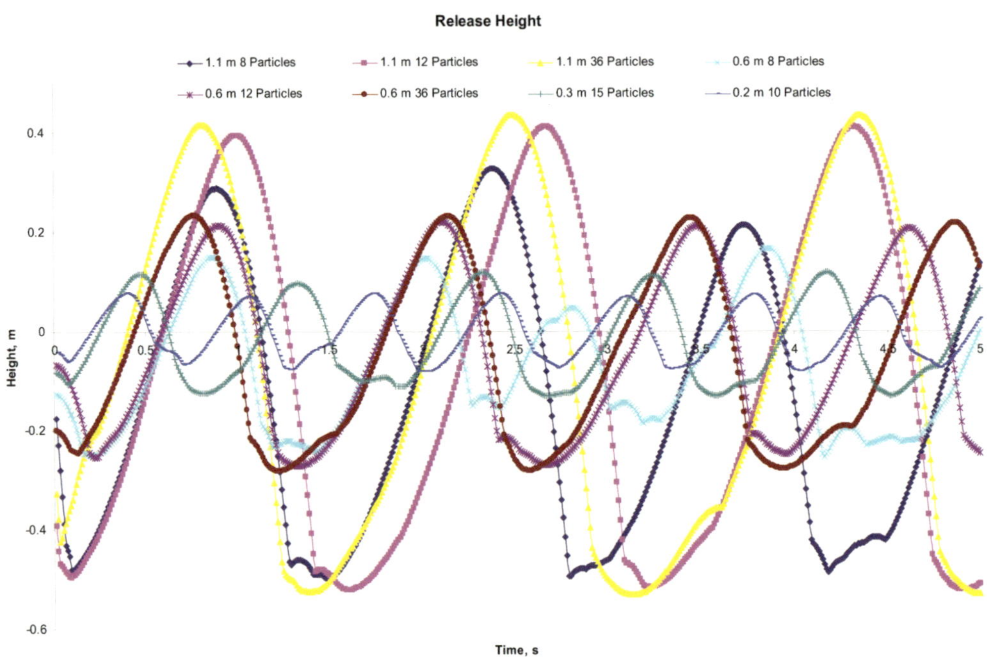

Fig. 2. Simulated release heights for different mills

4.1.3 Collision energy analysis

It is important to understand how particles accumulate and dissipate energy during abrasion. Figure 3 shows the energy gained by all particles in the 1.1 mill consisting of 20 particles for 5 s. Particles accumulate potential energy during the lift-up process. In the 1.1 m mill, particles are lifted higher than in the 0.6 m mill. Hence, particles in the 1.1 m mill gain and accumulate more energy. When the particles are released, they collide with other particles or the mill shell and transfer a portion of the accumulated energy to abrasion. The gain in the potential energy in the particles is confirmed by an inclined curve (from left to right) in Figure 3. When the particles are released from the lifters, the particles gain the velocity and the potential energy will be converted into the kinetic energy. The released particles hit other particles or the mill shell and use the kinetic energy to produce fragments (the disturbances in the vertical line). Due to the higher release height, in 1.1 m mill the particles have more available energy to release, hence, higher amount of potential energy will be imparted, thus higher kinetic energy.

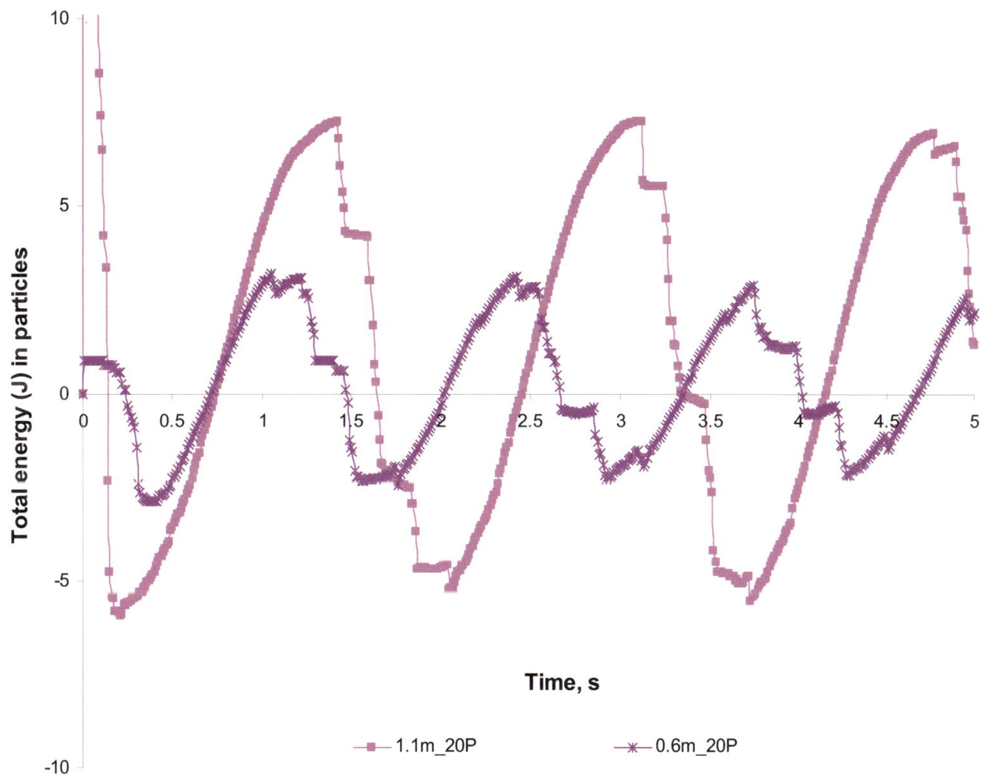

Fig. 3. Total energy gained by particles during tumbling

The same size distribution of particles in larger mills gains and dissipates higher energy than the smaller mills. However, the mill size is not the only factor which affects the gain in the energy for different size fraction of particles. The size fraction (size) of the particles and lifter size also influence the lift-up (release) height (Djordjevic et al., 2006) and hence, the collision energy.

4.1.4 Effect of mill diameter on abrasion

Three different mills (diameter 1.1, 0.6 and 0.3 m) were simulated with same parameters and throughput (eight particles) to investigate the effect of mill diameter on abrasion of particles. The mills were running at 2.53 rad/s (60% of critical speed for 1.1m mill). The same amount, size and mass of particles in different mills causes different "filled volume" in different sized mills. In smaller mills, the particle distribution is denser than larger mills with same number and size of particles. Table 6 shows the number of collisions in the mills. In the 0.3 m mill, due to less free space inside the mill, one can expect more collisions than in the other two cases. However, it seems from the data shown in Table 6 that it is not valid for 1.1 m mills. This maybe because of the very few particles in the larger mill, where the filled volume ratio (= total volume of particles/ volume of mill) was very low. The number of collisions is almost similar in the 0.6 and 1.1 m mills. This suggests that these two mills produce almost identical outputs for eight particles.

Mill diameter, m	0.3	0.6	1.1
Particles - all collisions, Hz	297	247	265

Table 6. Number of collisions for different mill sizes for the same mass of eight particles

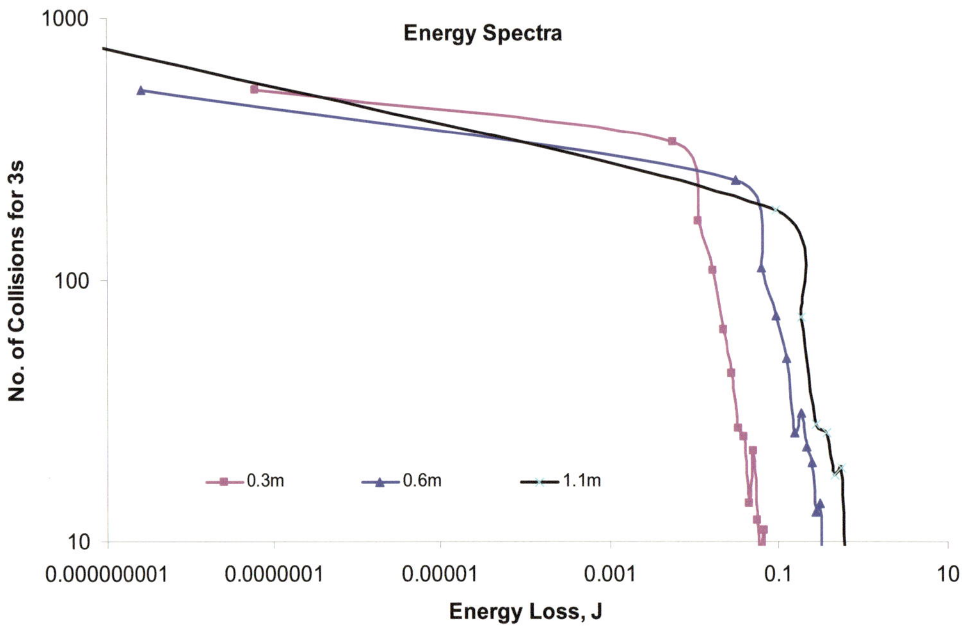

Fig. 4. Energy spectra for different mill diameters

Figure 4 shows the energy spectra for different sized mills (for the same parameters and particle loading conditions). The 0.3 m mill requires the least energy, among the three mills, to achieve the same amount of collisions. In short, the 0.3 m mill generates more collisions than the other two mills for the same expenditure of energy. Similarly, it can be observed that in the 0.3 m mill, particle-particle collisions are higher than in the other two types (Figure 5). This strengthens the hypothesis that a larger fraction of finer particles are generated in smaller mills than in the larger mills due to inter-particle collisions. The 0.3 m mill seems to be an energy efficient mill for disintegration of particles. However it generates more fragments than the others. As seen from Figure 5, the 0.6 m and 1.1 m mills have approximately same amount of inter-particle collisions with a noticeable difference in energy. For example, for 50 collisions, energy dissipated is approximately 0.15 J, 0.1 J and 1.5 J for 0.3 m, 0.6 m and 1.1 m mills, respectively. This shows, the 0.6 m mill is more energy efficient than the 1.1 m mill to achieve same number of collisions.

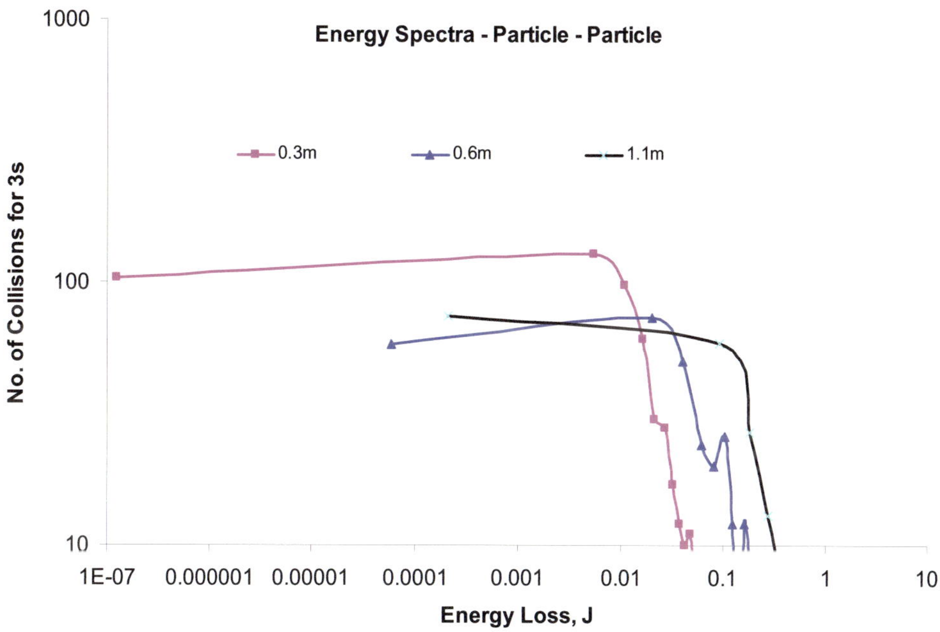

Fig. 5. Energy spectra (particle-particle) for different mill diameters

4.1.5 Effect of fraction of critical speed on tumbling

For similar diameter mills, different researchers have used different fractions of the critical speed of the mill. In this case, it is interesting to investigate the effect of critical speeds in tumbling parameters. The 1.1 m mill with 38 particles was chosen to study the effect of critical speeds in tumbling. Table 7 shows the number of collisions with respect to tumbling time at 60%, 70% and 75% of the critical speed. An increase in the tumbling speed causes a somewhat increase in the number of collisions, which is as expected. For 70% and 75% of the critical speed, there is not much difference in the number of collisions whereas at 60% and 70% of the critical speed there is noticeable increase in the number of collisions. This implies that a 5% variation in the rotating speed may not significantly influence the results. But in the long run (more than 30 s of simulation), going to 75% of the critical speed only causes a small increase in the number of collisions. This is further confirmed by Figure 6.

Critical speed, %	60	70	75
Particles - all collisions, Hz	1306	1501	1501

Table 7. Number of collisions for different critical speeds of the 1.1m mill

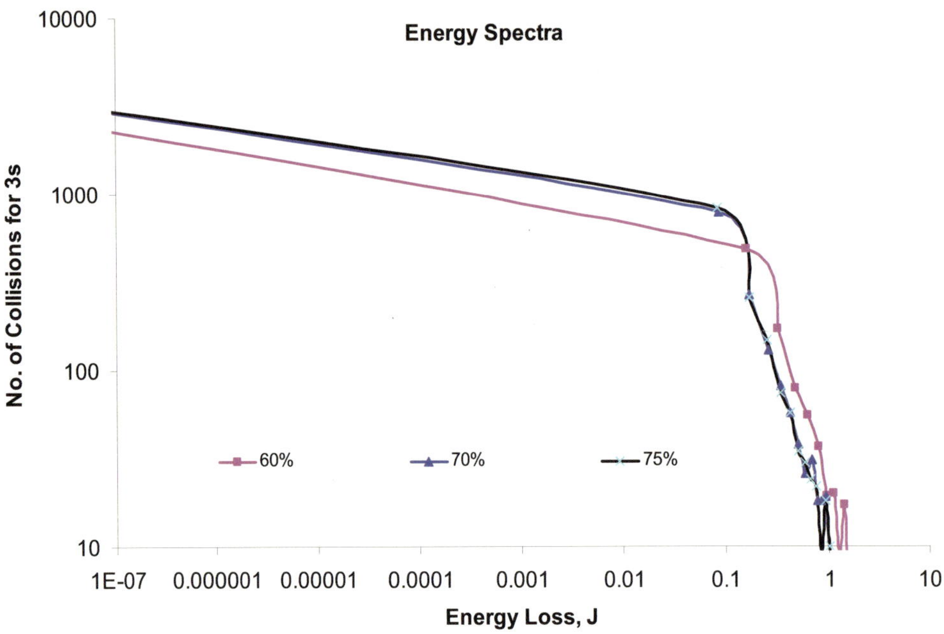

Fig. 6. Energy spectra observed for different rates and range of rotation.

Figure 6 shows the energy spectra at rates of rotation. At higher rotating speed relatively more energy is lost during particle collisions. The energy lost and the number of collisions is almost identical in the mill rotating at 70% and 75% of critical speed. However, the mill rotating at 60% of the critical speed shows relatively less collisions for almost same amount of energy loss. At lower speed (65% critical speed) the charges cannot reach the maximum height (compared to 70% and 75% critical speed), however the charges tumble effectively in the mill. This infers that the mill rotating at 70% and 75% of the critical speed will generate more fragments and is more suitable for incremental body break-age of the particles whereas lower rotating speeds are suitable for abrading the particles.

4.1.6 Effect of tumbling time

It has been observed that the effect of the tumbling time is pronounced. As the milling time increases, the cumulative number of collisions also increases. The number of collisions can be related to the abrasion of particles. However average number of collisions remains same. In the experiment, the particle size distribution curves show the generation of more fine material as the grinding time increases. Due to a large number of collisions in the beginning of the simulation, higher energies are lost at the initial stage of the tumbling. Once the mill reaches steady state, the rate of energy losses also become uniform and reaches a steady state condition. The effect of grinding time in collisions can be verified by the energy spectra curves observed at different tumbling times.

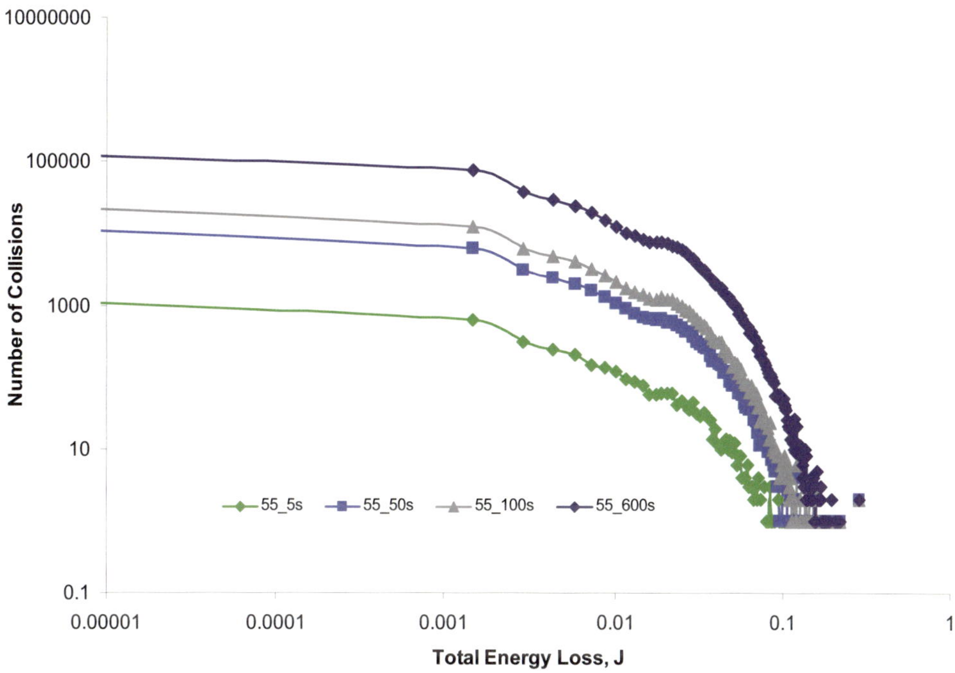

Fig. 7. Number of collisions versus total energy loss for -55.5+37.5 class particles for different grinding time (Xs_in the legend shows the milling time).

The frequency distributions of the energy losses can be determined from the individual collision events. These collision energy spectra provide an opportunity for improved understanding of the various contributions to the overall energy dissipation within the mill. Knowledge of effect of mill parameters on the energy spectra will allow in selecting useful attributes to increase the efficiency of comminution. Figure 7 shows the collision - energy loss relationship for the abrasion at different time. The energy spectra distribution in the graph shows a similar trend. At short tumbling time, few collisions occur, so less energy loss. As the grinding duration increases, the number of collisions also increases. The maximum energy dissipation is relatively larger for more tumbling time than for less tumbling time. A large number of collisions are observed at the lower energies. Based on a hypothesis - number of collisions increases the production of fragments - the 600 seconds tumbling time produces more fragments with fines than the 5 seconds tumbling time with an identical size fraction. Due to a large number of collisions in the beginning of the simulation, higher energies are lost at the initial stage of the tumbling.

4.1.7 Effect of particle size

Experimental investigations showed that the effect of particle size in abrasion is not so influential (Unpublished report). It has been reported that the changes in particle size for the same tumbling time of 600 seconds tend to produce similar curves for the smaller sizes.

In the DEM simulations, the number of particles in the mill was varied to represent different particle sizes for the same total mass. For example, 3 kg of total mass can be represented by the size classes -55+37.5, -37.5+26.5, -26.5+22.4 and -22.4+16 with 24, 72, 157 and 334 particles, respectively.

It is evident from the literature that the particle–particle collisions show different failure mechanism than the particle–wall collisions and the particle velocity also govern the severity of particle failures (Khanal et al, 2007). Table 8 shows that the fewer inter-particle collisions occur in the tumbling mill than the particle-mill shell collisions. Hence, majority of particle breakage occurs due to particle-mill shell collisions. As the number of particles increases in the mill, the number of collisions (total and inter-particle) also increases, which are shown by -22.4+16 mm size class particles. More particles show more collisions with a large number of inter-particle collisions. This demonstrates that large number of particles in a mill increases particle–particle collisions and a reasonably high amount of dust will be produced. This was observed experimentally by Banini, (2000); Loveday and Naidoo, (1997). The rock overcrowding at the toe of the mill also enhances the inter-particle collisions.

Size class (mm)	Total collisions (MHz)	Particle-particle collisions (MHz)
55-37.5	10.17	33.04
37.5-26.5	32.52	16.22
26.5-22.4	80.32	56.36
22.4-16	186.24	130.67

Table 8. Total and particle-particle collision for different size classes for 600 seconds

Figure 8 shows the relationship between number of collisions and total collision energy loss for 5 seconds. The four curves follow a similar trend in dissipating the collision energy. Large number of collisions occurs at the lower energy levels. The -55+37.5, -37.5+26.5, -26.5+22.4 size classes require more energy to achieve the same number of collisions as occurred in -22.4+16 size class. The density of the particles (population) in -22.4+16 size class is higher than in the other three classes, so, the number of collisions in former is higher than in the latter. Therefore, it can be suggested that the particle size also influences the generation of fines. Material behaviour changes from elastic to plastic as particle becomes smaller (Rumpf, 1995). It is a very difficult and high energy process to break (or abrade) the tiny particles as those particles may have very few flaws present in them compared with large particles.

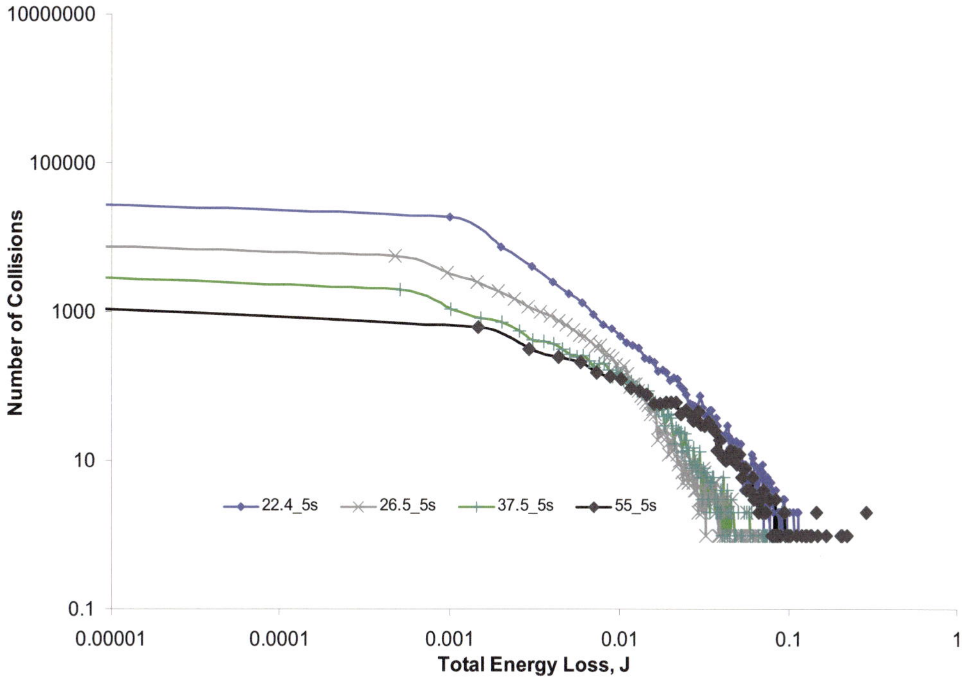

Fig. 8. Collisions versus total energy loss for different size classes after 5 seconds of mill operation.

4.1.8 Effect of particle modulus

The particle size distribution of the progeny depends upon the type of parent materials. In other words, the variation in the material properties causes the difference in progeny particle size distributions. The experimental progeny particles are related to the simulated number of collisions as, in simulations number of collisions causes the generation of daughter particles. It can be expected that the larger modulus of particles (stiff particles) generate relatively more collisions than the lower modulus of particles (soft particles).

Since the modulus is a measure of the stiffness of a given material, four different shear moduli (0.1, 1, 10 and 100 MPa) were chosen to represent different ore types ranging from very soft to moderate as hard. Each ore type is simulated for exactly the same conditions except for the change in shear modulus. Figure 9 shows the number of collisions versus total energy loss for different types of particles. The four curves follow similar trends in dissipating the collision energy. Large numbers of collisions occur at the lower energy levels. The low shear modulus particles require more energy to achieve the same number of collisions as occurred with high modulus particles. The figure shows that as the modulus of particles increases the number of collisions also increases. In other words, at the same amount of energy, the higher shear modulus of particles generates more collisions than the lower modulus of particles. This suggests that stiff particles should be more energy efficient in generating fragments than the less stiff particles.

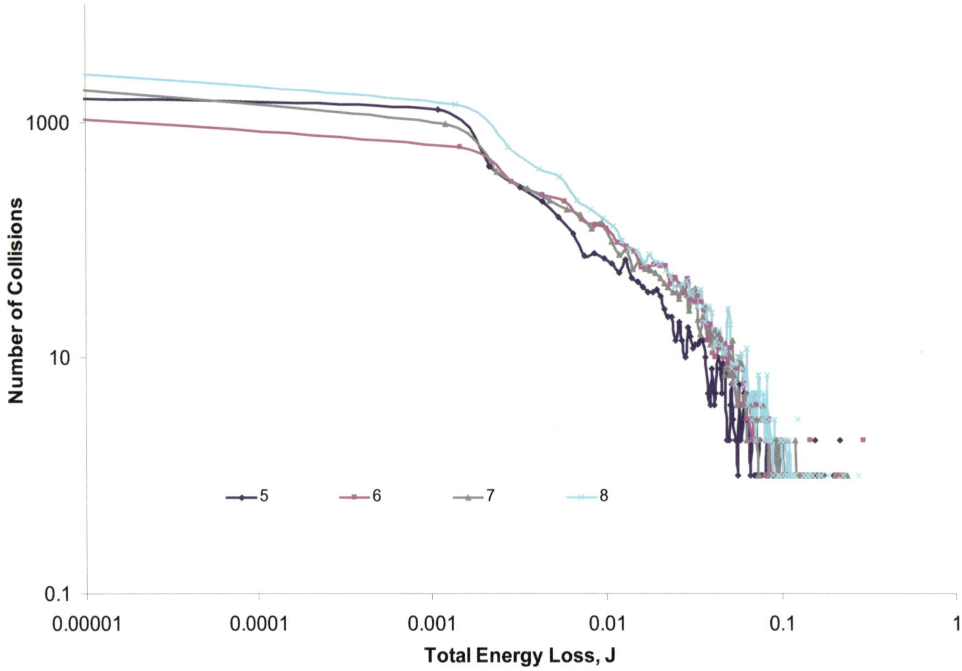

Fig. 9. Collisions versus total energy loss for different particles (5, 6, 7 and 8 represent shear modulus of 1e5, 1e6, 1e7 and 1e8 Pa respectively).

4.2 Non-spherical particles

4.2.1 Particle trajectories

Figure 10 shows the particle trajectories in a 1.1m mill for different non-spherical particles. Particles follow the circumference of the mill up to a maximum height the lifter can take them. Then, due to gravity, the particles fall randomly within the mill. However, once the mill reaches a steady-state condition, the lifters try to throw the particles as far as possible in a projectile trajectory. Depending on the population (density) of particles inside the mill, the fallen particles hit either ball or shell. For the optimum condition, the particles do not collide with each other before hitting the mill shell. But if the particles are overcrowded, notable inter-particle collisions will occur and finer particles will be generated. Hence, particle-particle interaction does not appear to occur until a critical number of particles. An increase in the mill load results in an increase in wear rate of the particles. The shorter trajectories for the same particle in the graphs are due to the data extraction at the unsteady-state condition (i.e., before 2.3 s of simulation). As the centre of mass of smaller particles is closer to the mill shell, they are lifted a little higher than the larger particles.

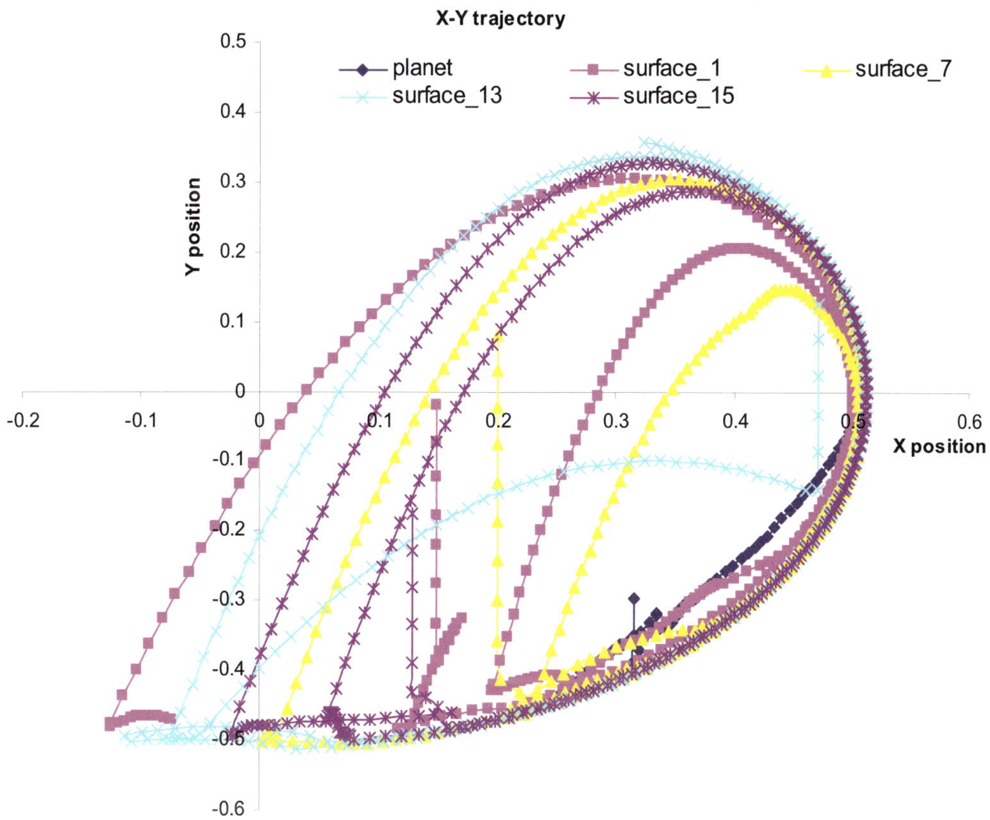

Fig. 10. Particle trajectories for particles of different surface "roughness" (mill speed = 2.53 rad/s, 60% of the critical speed)

4.2.2 Collision energy analysis

Figure 11 shows the loss of collision energy for non - spherical particles with respect to tumbling time. Due to a large number of collisions in the beginning of the simulation, higher energies are lost at the initial stage of the tumbling. Once the mill reaches steady-state condition, the rate of energy losses also become uniform and reach a steady-state condition. The less spherical-like surfaces experience a larger number of particle-wall collisions than the more spherical-like particles; as a result, the former loses higher energies in collisions than the latter. This is due to the fact that rougher surfaces have more surface area and contact points than the spherical particles. The mill reaches steady-state after 2.3 s of tumbling time. The initial peaks observed at 0.01 s are simulation artefacts during filling of particles in the mill.

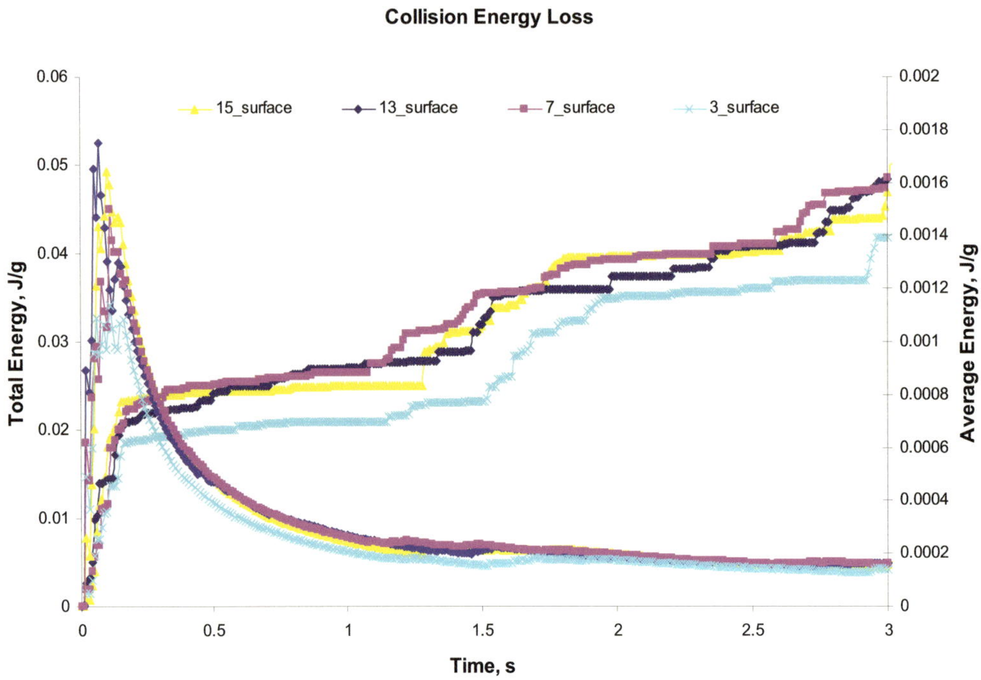

Fig. 11. Total collision energy loss vs. time (mill speed¼2.53 rad=s)

4.2.3 Energy spectra analysis

The frequency distributions of the energy losses can be determined from the individual collision events. These collision energy spectra provide an opportunity to better understand the various contributions to the overall energy dissipation within the mill (Cleary and Morrison, 2004). Knowledge of the effect of mill parameters on the energy spectra will allow selecting useful attributes to increase the efficiency of comminution.

Figure 12 shows the number of collisions with respect to (a) total collision energy loss, (b) normal collision energy loss, and (c) tangential collision energy loss for 3 s of tumbling. Each graph compares the energy losses for different types of particles. The energy spectra distribution in these three graphs shows a similar trend. The maximum energy dissipation is larger for rough particles than smoother particles. The smoother particles experience fewer collisions than the rougher particles, although the total energy dissipation is somewhat similar. A large number of collisions are observed at the lower energies. As noted by Cleary and Morrison (2004), the introduction of additional fines generates more low energy collisions; the rougher particles (having more humps, e.g., 13 surface total, 15 surface total) experience more collisions than the smoother particles at the low energy levels. So, it can be analogized that the rough surfaces or humps are in some way similar to the additional fine particles tumbled in the mill. The input energy per unit mass of particle is an important factor to assess the probability of breakage. Therefore, the humps of the particles may achieve large input energy per unit mass in a collision (sandwiched between particles) with other main particles and are much more likely to be broken because the stress is concentrated in a small zone. Figure 12(b) shows that during collisions, normal energy dissipation is more responsible than the tangential energy dissipation (Figure 12(c)) for abrasion of rougher particles.

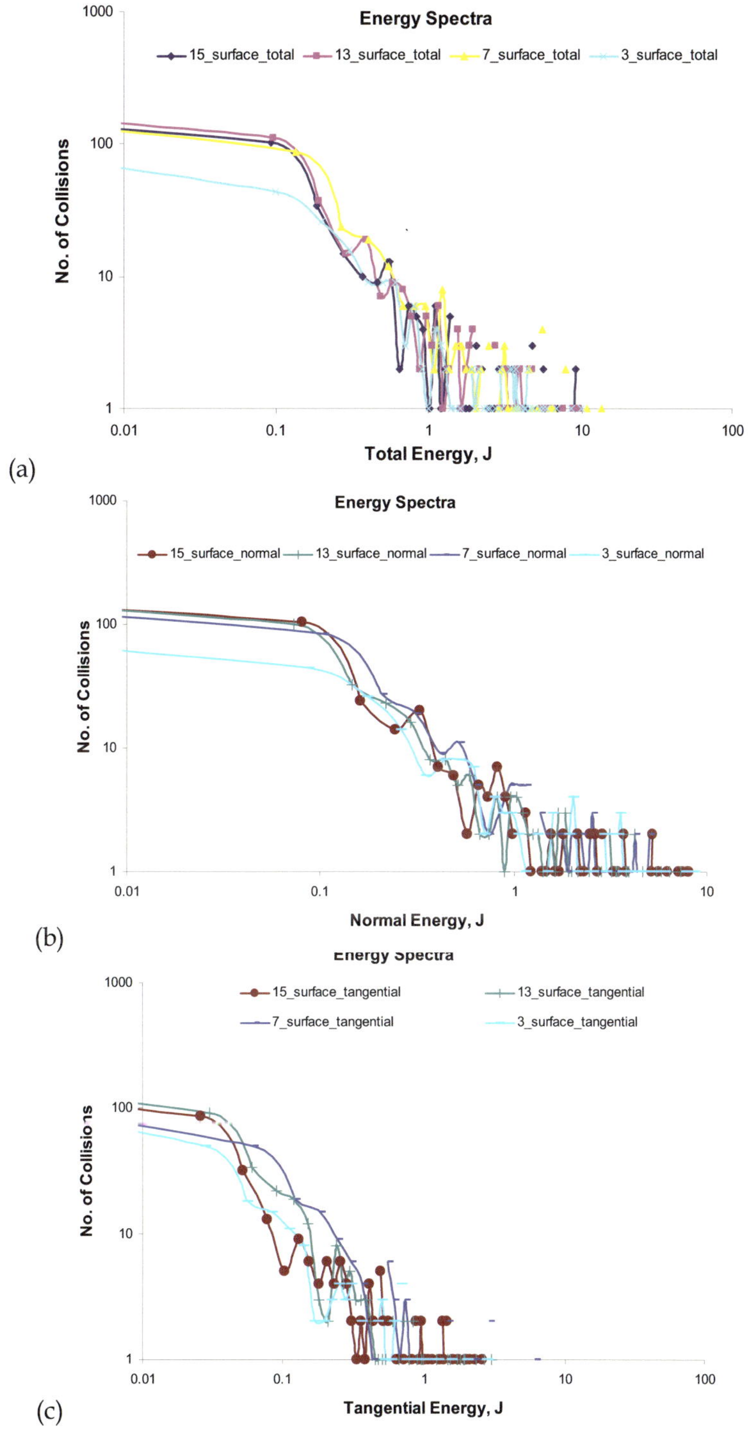

Fig. 12. Number of collisions vs. (a) total energy loss, (b) normal energy loss, and (c) tangential energy loss (mill speed¼2.53 rad=s)

4.3 Testing of hypothesis

Revisiting the hypothesis, the number of collisions increases the mass loss from the parent particles. The experimental mass loss has been compared with the simulated collision energy to test the hypothesis, Figure 13. As mentioned earlier, the tumbling time required to conduct the abrasion test was obtained from the experiments. The abrasion experimental mass loss was compared with the simulated number of collisions, shown in Figure 13.

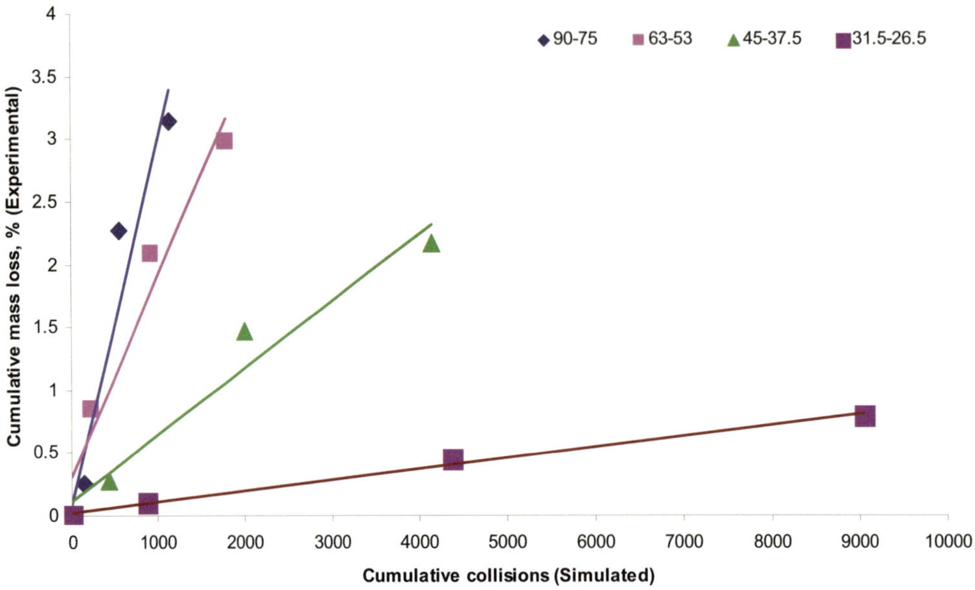

Fig. 13. Experimental mass loss as a function of number of collisions.

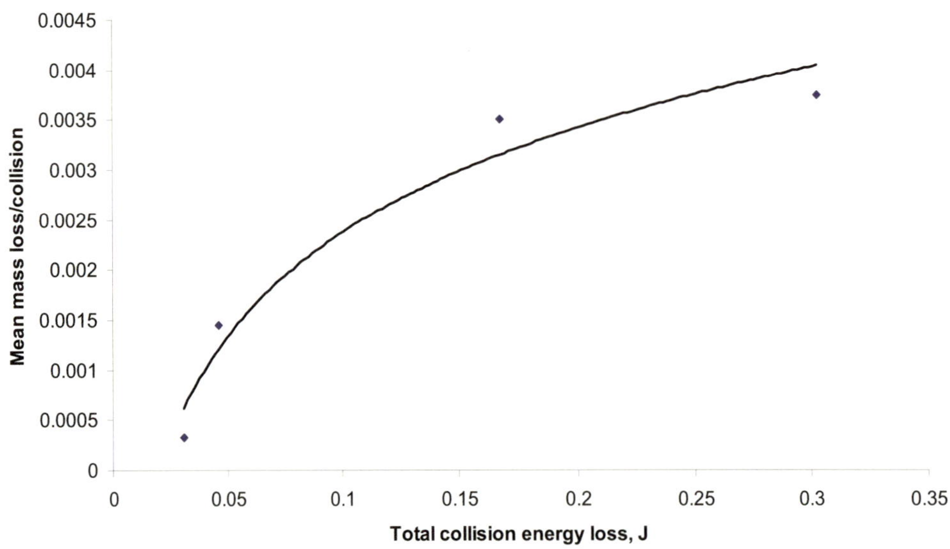

Fig. 14. Mean mass loss per collision (experimental) with respect to the total collision energy estimated from DEM in each size fraction.

The preliminary observations show that as the number of collisions increases, the mass loss from the particles also increases. This indicates that the hypothesis being tested is reasonable. The trend lines in Figure 14 show a very good fit except for 45-37.5 class. Smaller particles lose less mass compared to the larger particles for the same number of collisions. However, due to the limited data points on the graph, it is difficult to relate the nature and degree of dependency of mass loss on the number of collisions for different sized particles.

5. Conclusions

The simulated milling has been carried out in a particle tumbling range (time) which is well below the starting energy required for body breakage of particles. Although the preliminary study shows that the assumed hypothesis (the number of collisions increases the mass loss from the parent particles) is valid, many more experiments have to be carried out to find the nature and degree of dependency of mass loss on abrasion on the number of collisions.

For the less number of particles (considered here for abrasion) the particle-wall collisions are the major cause of particle breakage in the mill as the number of inter-particle collisions are much less compared to the particle-wall collisions. The energy (gain and loss) and energy spectra analysis of particles provided an insight into the dissipation of energies during collision. The force imparted into the particles could be a parameter which depends on the material and the loading process. Hence, further experiments and simulations need to be carried out to study the mechanism of force/energy dissipation during collisions.

The paper also discussed the effect of various abrasion process parameters to investigate their effect in abrasion mechanism of particles. It has been noted that the grinding time affects the comminution of particles. As expected, longer grinding time produces more fragments along with more fines. The size and stiffness of particles also affect the degree of abrasion. For the same mass, smaller particles show more collisions and hence, high degree of abrasion. However, it has been realized that experimental observations are required to validate the effect of size and stiffness of particles in abrasion.

Based on this hypothesis (the number of collisions increases, the mass loss from the particles also increases), it can be concluded that the more irregular particles lose their irregularities (humps and asperities) faster than the more rounded particles. It was observed that as the asperities of the particle decrease, the rate of abrasion decreases. The particle-wall collision is the major cause of particle failure in the mill within the optimum number of particles.

The energy analysis provided insight into the dissipation of energies during collision During particle-mill shell or particle-particle collision, the asperities of the particles first come in contact with the colliding partner. These asperities will act as stress concentrators on the particles. Based on the energy (gain and loss) and energy spectra analysis of the particles, it was found that the normal collision force is larger than the tangential collision force, and hence the former may be more responsible for the abrasion of particles, i.e., the removal of the asperities of the particles.

6. References

Banini, G.A., 2000. An Integrated Description of Rock Break-age in Comminution Machines, in Julius Kruttschnitt Mineral Research Centre, vol. Ph.D., niversity of Queensland, Brisbane (unpublished).

Cleary, P., Morrison, R., 2004. The Role of Advanced DEM Based Modelling Tools in Increasing Comminution Energy Efficiency. In: Presented at Green processing.

Djordjevic, N., Morrison, R., Loveday, B., Cleary, P., 2006. Modelling comminution patterns within a pilot scale AG/SAG mill. Minerals Engineering 19, 1505- 1516.

EDEM, 2006. DEM Solutions, Edinburgh.

H. Rumpf,Particle Technology, (Translated by F. A. Bull) Chapman and Hall, London (1995).

Khanal, M, Morrison, R, 2008 Discrete element method study of abrasion, Minerals Engineering 21, 751-760

Khanal, M, Schubert, W, Tomas, J 2007, Discrete Element Simulation of bed comminution, Minerals Engineering, Minerals Engineering 20, 179-187

Loveday, B., Morrison, R., Henry, G., Naidoo, U., 2006. An investigation of rock abrasion and break-age in a pilot-scale AG/SAG Mill. In: Presented at SAG 2006, Vancouver, B.C., Canada.

Loveday, B.K., Dong, H., 2000. Optimisation of autogenous grinding. Minerals Engineering 13, pp. 1341-134.

Loveday, B.K., Naidoo, D., 1997. Rock abrasion in autogneous milling. Minerals Engineering 10, 603-612.

Loveday, B.K., Whiten, W.J., 2002. Application of a rock abrasion model to pilotplant and plant data for fully and semi-autogenous grinding. Transaction Institute of Mining and Metallurgy 307, C39-C43.

Morrison, R.D., Cleary, P.W., 2004. Using DEM to model ore break-age within a pilot scale SAG mill. Minerals Engineering 17, 1117-1124.

Solidworks, 2006. Solidworks Inc., SP 4.1 ed.

5

Low Impact Velocity Wastage in FBCs – Experimental Results and Comparison Between Abrasion and Erosion Theories

J. G. Chacon-Nava[1,*], F. Almeraya-Calderon[1], A. Martinez-Villafañe[1] and M. M. Stack[2]
[1]*Department of Integrity and Design of Composite Materials Advanced Materials Research Center CIMAV, Chihuahua, Chih.,*
[2]*Department of Mechanical and Aerospace Engineering University of Strathclyde, Glasgow, Scotland,*
[1]*Mexico*
[2]*UK*

1. Introduction

The use of technologies related to combustion of coal in fluidized bed combustors (FBCs) present attractive advantages over conventional pulverized coal units. Some of the outstanding characteristics are: excellent heat transfer, low emission of contaminants, good combustion efficiencies and good fuel flexibility. However, FBC units can suffer materials deterioration due to particle interaction of solid particles with the heat transfer tubes immersed on the bed (Hou, 2004, Oka, 2004, Rademarkers et al., 1990). Among other issues, some of the most important factors believed to cause wear problems are: the motion of slowly but relatively coarse particles, particles loaded onto the surface by other particles, erosion by relatively fast-moving particles associated with bubbles, and abrasion by blocks of particles thrown into the surface by bubble collapse. Thus, erosion or abrasion processes can occur by a variety of causes. For the case of particle movement against in-bed surfaces, it has been suggested that there is no difference in the ability to cause degradation between solid particle erosion and low stress three body abrasion, and distinctions between the two forms of wear should not to be made (Levy, 1987).

1.1 The most commons types of FBCs

On applications such as steam and power generation, the most important types of FBCs are: 1) the atmospheric fluidized-bed combustor (AFBC). The superficial air velocity is between 1 and 3 m s^{-1}, to give a "bubbling bed"(Highley & Kaye, 1983); 2) The pressurized fluidized-bed combustor (PFBC). Here, the unit is operated at elevated pressure (from 6 up to 40 bar), and the main purpose is to expand the combustion products in a gas turbine to generate

* Corresponding Author

electricity through steam rising. Therefore, a higher efficiency of electricity generation is possible than that from either a gas or steam turbine plant alone (Howard, 1989); 3) the circulating fluidized-bed combustor (CFBC). In this system, the velocity of the fluidizing gas is significantly higher, being of a typical value about 5 m s^{-1} to 10 m s^{-1} than in the two previous systems. Fig. 1 shows a schematic diagram of the various FBC´S systems (Rademarkers & Ketunen, 1986).

1.2 Wastage problems in FBCs

Earlier studies on material behavior in FBCs suggested that wastage by particulate erosion did not represent a potential problem. A report considerer very unlikely that wastage would be a serious problem as the particle impact velocities are low, generally less than 5 m s^{-1}, (Mezko, 1977). Another study confirmed the above observations, since in their FB unit they did not found evidence of wastage on the in-bed tubes (Beacham & Marchall, 1979). However, despite the general good signs expressed above, since the early eighties, material wastage has been a recurrent problem in bubbling fluidized-bed combustors (AFBC and PFBC) throughout the world. For instance, wear rates of about 1 mm per 1000 h for in-bed tubes have been reported in a Chinese unit (Zhang, 1980). In the USA, observations of wastage of the in-bed tubes in several combustors have been reported (Kantesaria & Marchall, 1983, Montrone, 1983). In the UK, high wastage has been reported from the Grimethorpe plant, where wastages rates up to 1.7 mm per 1000 h were recorded on evaporator tubes (Anderson et al. 1987).

1.3 Wear characteristics

It is not clear to what extent materials wastage can be attributed to mechanical phenomena such as erosion and/or abrasion by the fluidized bed particles (Stringer & Wright, 1987). Some of the most important factors believed to cause problems are: a) erosion by relatively slowly moving, but relatively coarse particles; b) wear by particles loaded onto the surface by other particles; c) erosion by relatively fast-moving particles associated with bubbles; d) intrinsic fast particles in the bed (apart from the fast particles in the bubble wakes); e) erosion by fast moving particles immediately above the bed in the splash zone; f) erosion or abrasion associated with long range patterns; g) erosion induced by in-bed jets; h) abrasion by blocks of particles thrown into the surface by bubble collapse and i) wastage induced by the presence of geometrical irregularities

The wastage modes within a bed are not well understood because some of the above causes can be more closely related to a purely erosion process, for example c), e), f) and g), whereas others appears to be more related to a three-body abrasion process, for example b) and h). A general representation of the dense and bubble phase on an in-bed surface is presented in Fig. 2 (Janson, 1985)

Frames d) to f) are of interest regarding particle –tube interactions; in d) and e), changes in emulsion phase density can occur as conditions change in the fluidized bed, for example, fluidization behavior. In e), packing of the emulsion phase against an in-bed tube can be seen. Frame f) shows the case where a bubble. With a limited number of particles, is present at the bottom of a tube. For in-bed tubes, it is generally accepted that particle impact velocities range from 1 m s^{-1}, to 5 m s^{-1}. However, experimental determinations of particle velocities are difficult and rely on cold model studies. For instance, it has been reported that the average particle velocity was about 70% of the superficial velocity (Vs), but some particles had velocities as high as five times the values of Vs (Boiarski, 1978). Another study found that on bubble arrival at the tube surface, the particle velocity increases rapidly and streak across the tube surface, at velocities up to 5.6 m s^{-1}(Peeler & Whitehead, 1982). Another work found that particles do not move independently but as aggregates, and reported expressions giving the particle velocity, Vp, as function of the gas superficial velocity and impact angle, in the form (Tsutsumi et al., 1989):

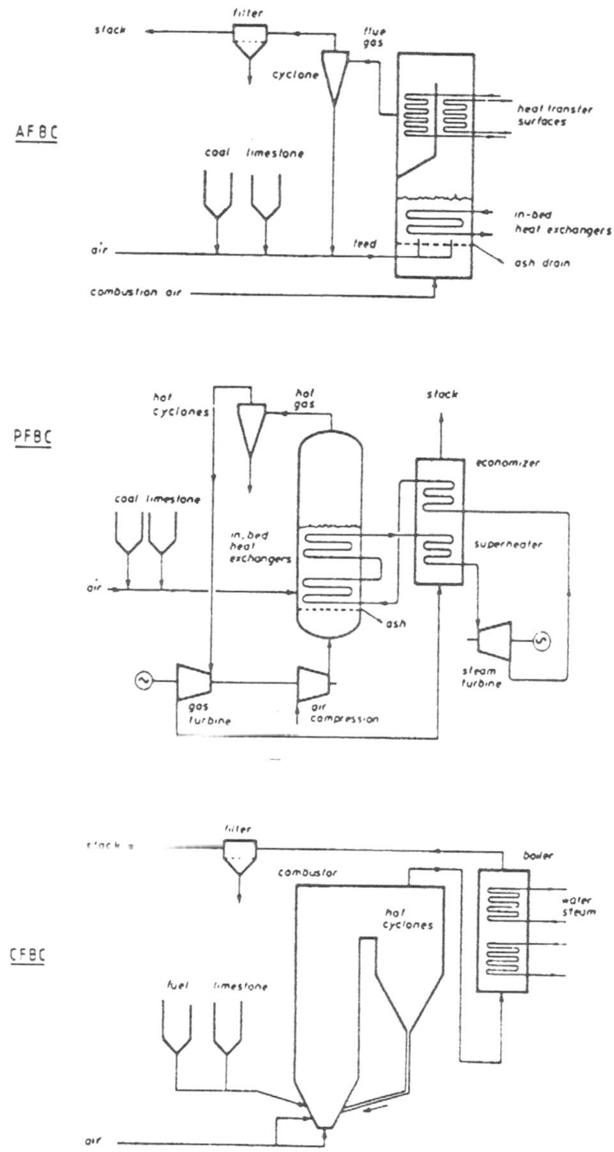

Fig. 1. Most common types of FBCs.

$$Vp = 3.24\ Vs^{0.73} \quad \text{(at 45}^{0}\text{ around the tube)} \qquad (1)$$

$$Vp = 1.99\ Vs^{0.89} \quad \text{(at normal impact angle)} \qquad (2)$$

As Vs is expressed in m s^{-1} the first expression (valid at shallow impact angles) predicts a particle impact velocity of 3.24 m s^{-1} at a superficial velocity of 1 m s^{-1}. However, since many FBC units typically operate at Vs about 2.5 m s^{-1}, particle velocities of about 6.2 m s^{-1} can be obtained.

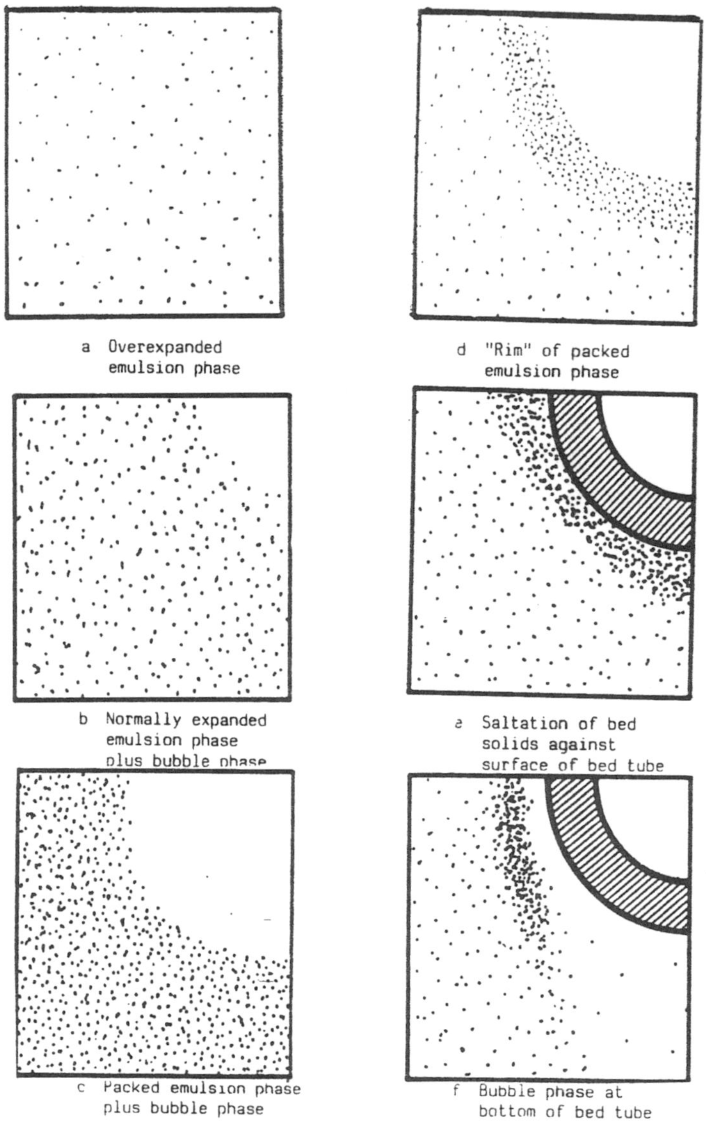

Fig. 2. A representation of variations in solid density in fluidized-bed bubble and emulsion phases.

1.4 Location of maximum wear

The maximum wear is normally located on in-bed tubes on the bottom half portion of the tubes, and appears in two patterns. In one, the wear is a maximum at positions about 20^0 - 30^0 on either side of the bottom, while wear is small or zero at the bottom. The second pattern has the maximum wear at the bottom of the tube, decreasing to zero typically at about 45^0. These patterns have been called "Class A" and "Class B" respectively (Stringer & Wright, 1987), and have been observed in cold simulations as well as in practical units (Anderson et al., 1987, Parkinson et al., 1985, Tsutsumi et al, 1989, Wang et al., 1992). It is interesting to note that Class A is related to the angle of impact corresponding to maximum wear in erosion of ductile metals. However, cold model studies speculates that, erosion is a maximum at normal impact angle, while abrasion is a maximum at about 35^0 on either side of the tube bottom (Wheeldon, 1990). Class B is more typical of the so-called brittle erosion. Fig. 3 shows both patterns which have been observed in real combustors.

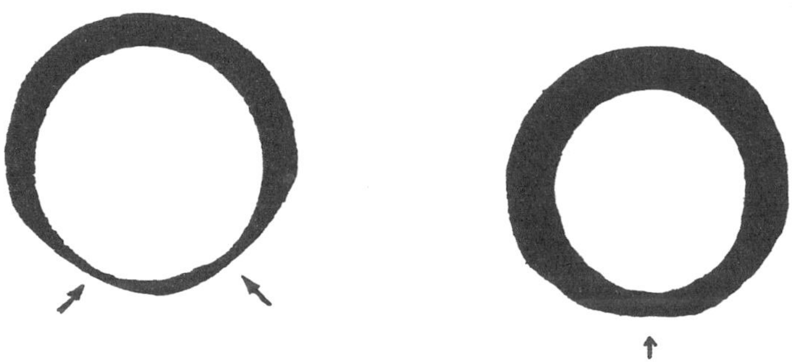

(a) (b)

Fig. 3. Wastage patterns found in FBCs indicating with arrows the maximum location of wear: a) Class A, b) Class B.

As was mentioned before, erosion or abrasion (particularly three-body abrasion) on in-bed tubes can occur by several causes. To date it is not clear to what extent materials wastage can be attributed to phenomena such as abrasion or erosion by the fluidized bed particles, and the predominance between erosion or abrasion as main forms of mechanical damage remains as an area of discussion. The aim of this work is to assess the effect of impact angle and mode of wear in terms of the observed morphologies of the exposed surfaces and 1) a dominant abrasion process derived from a simulated fluidized bed environment and Rabinowicz theory and 2) an erosion process, using Finnie´s erosion theory.

2. Experimental procedures

2.1 Materials

Cylindrical specimens with typical dimensions of about 6 mm diameter and 24 mm length made from mild steel (in normalized condition) and 310 stainless steel (solution treated) were used. Table 1 shows the chemical composition (%wt) for the steels.

	C	S	P	Si	Mn	Cr	Ni	Mo	Fe
Mild steel	0.22	0.03	0.04	0.35	0.72	-	-	-	Bal.
310 SS	0.15	0.025	0.03	1.5	2.0	25	19	-	Bal.

Table 1. Chemical compositions (%wt) of the steels tested

Before exposure, the specimens were ground progressively to a final surface finish with 800 grit SiC paper using water as coolant, immediately rinsed in methanol, degreased with acetone and dried in a stream of hot air. The extent of damage to specimens exposed to the erosion rig was determined by weight change per unit area measurements using a Sartorious microbalance, with a resolution of 10^{-5} g.

2.2 Impact angle measurements

To assess the effect of impact angle on selected specimens after tests, thickness loss measurements were carried out using a profilometer system. Basically, the system consist of i) a specimen jig and its movement unit; ii) a stylus coupled with a LVDT transducer and iii) the acquisition unit. The specimen is positioned in the jig which is rotated horizontally about its axis by a stepper motor coupled to a gear box. To perform a measurement, a stylus is brought into contact with the specimen surface in a continuous mode, and any vertical displacement is taken up by movement of the core of a LVDT transducer with a linear displacement range of ± 1 mm and a reproducibility of 1 µm. Before the start of a run, the stylus was positioned on the area (trailing edge) that was not exposed to the particles in the bed environment, using this as a base line.

2.3 The fluidized-bed rig

Experiments were carried out in a fluidized bed (FB) rig, which basically consist of a) a fluidized-bed chamber containing approximately 40 % vol. of particles during a test, b) a specimen holder system and c) a heating system. Fig 4 shows a schematic of the apparatus used. This consists of a light fluidized-bed of particles in which cylindrical specimens are rotated in the vertical plane into and out of the bed. For each temperature, a fluidizing velocity of 1.3 X Umf (where Umf is the minimum fluidization velocity) was used. Depending upon the angular velocity chosen, the linear velocity of the specimens relative to the particles is achieved.

2.3.1 Experimental conditions

The experimental conditions used were the following: exposure temperatures from 100^0C up to 600^0C, and each test last 24 hours; air as oxidant gas; the particles used were relatively angular alumina particles of 560 µm average size at impact velocities ranging from 1 m s^{-1} to 4.5 m $s.^{-1}$. Owing to degradation, the particles were replaced at regular intervals. Morphological examinations of wear scars were carried out using an AMRAY scanning electron microscope, linked with an EDX unit. The aim here was to assess the main trends of the effects of temperature and impact velocity and to characterize the extent of degradation of the specimens.

3. Results and analysis

3.1 Wastage as a function of temperature and impact velocity

3.1.1 Mild steel

Fig. 5 shows the behaviour of mild steel. At the lowest velocity and below 250ºC, no weight losses were recorded. Further increases in temperature resulted in small weight gains (negative scale). At 2 m s^{-1} and temperatures above 500ºC, this trend changed, and small weight losses were recorded. In the velocity range of 2.5 m s^{-1} to 4 m s^{-1}, a wastage peak was observed. This peak occurred at 300ºC, for impact velocities up to 3.5 m s^{-1}, but between 300ºC and 350ºC at 4 m s^{-1}. The weight losses had a minimum at 450ºC, and, above this temperature, significant increments in weight loss were recorded. At 4.5 m s^{-1}, no wastage peak was observed only a continuous increase in weight loss with temperature.

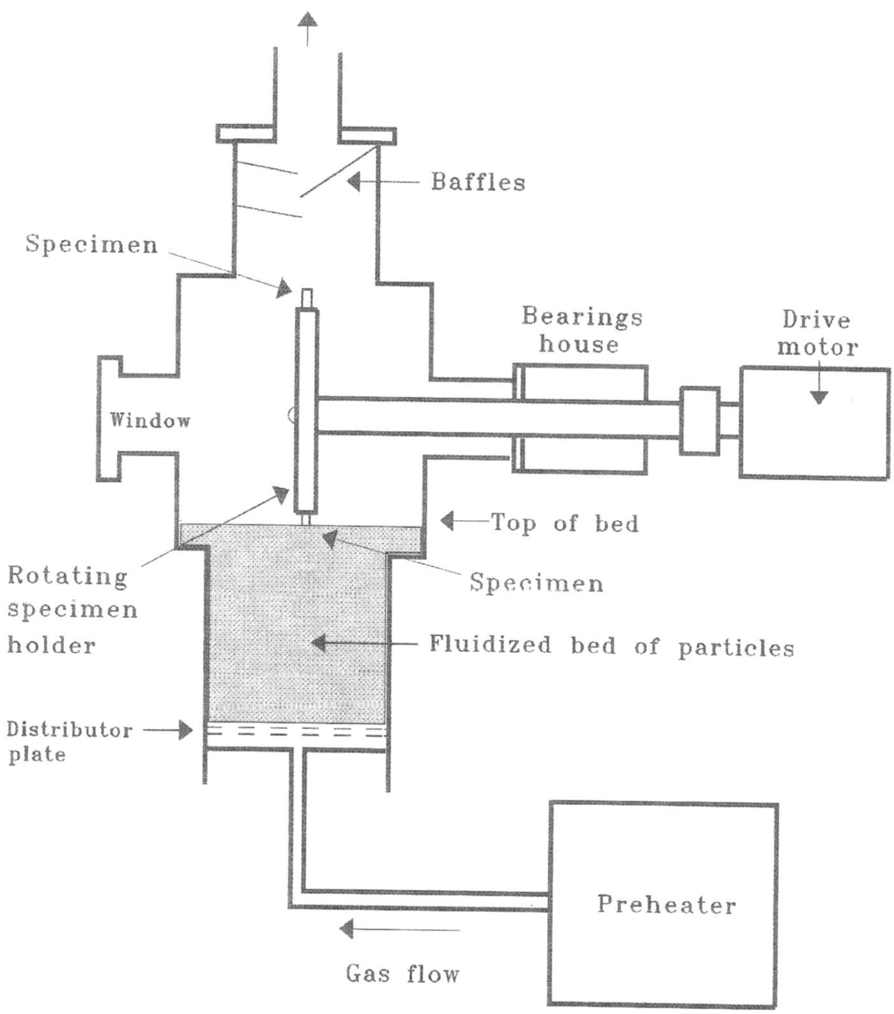

Fig. 4. Schematic diagram of the fluidized-bed apparatus used.

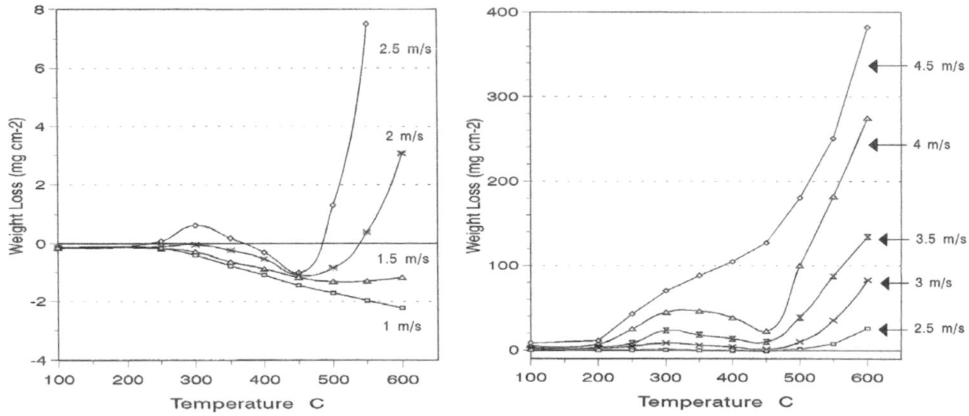

Fig. 5. Weight change as a function of temperature and specimen velocity for mild steel exposed in the FB rig with the 560 μm alumina particles for 24 h.

Fig. 6 shows the morphology after exposure at 250^{0}C and 1.5 m s^{-1}. Here, the surface consisted mainly of compacted erodent (and erodent debris). X-ray mapping on the surface gave evidence for this. At 300^{0}C and 2.5 m s^{-1}, a rather different morphology was observed: the surface had a polished appearance, and, at higher magnification, a thin (apparently less than 1 μm thick) scale was observed, Fig. 7. At 450^{0}C and 2.5 m s^{-1}, the surface had a rippled appearance, Fig. 8, whereas at 600^{0}C and 1.5 m s^{-1}, the surface was again rippled, and scale had apparently spalled from some areas, Fig. 9 (a). At 2.5 m s^{-1}, surface ripples were still noted, but, now, cracks in the surface scale were clearly seen, Fig. 9 (b). With further increase in velocity to 4.5 m s^{-1}, surface ripples were no longer observed but a dark polished surface was noted. At higher magnification, surface cracks were observed, Fig. 9 (c).

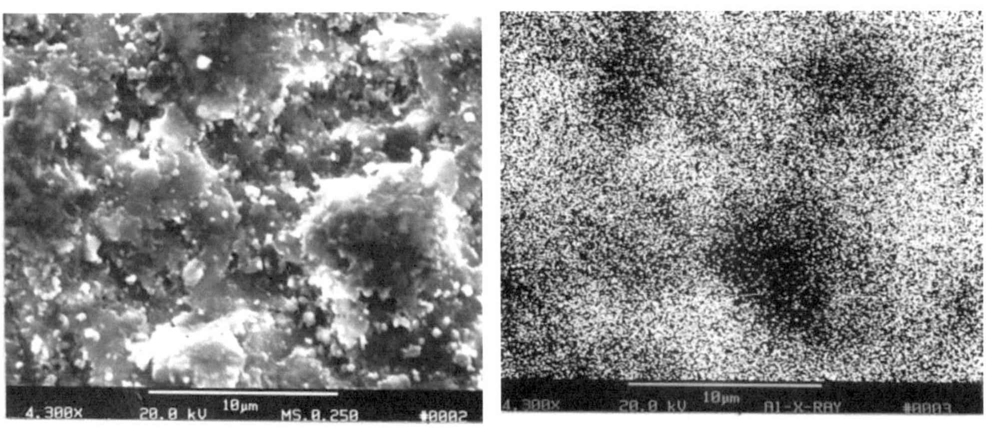

(a) (b)

Fig. 6. Scanning electron micrograph of mild steel exposed in the FB rig at 250^{0}C and 1.5 m s^{-1} showing a) the surface morphology and b) X-ray map of aluminum in a)

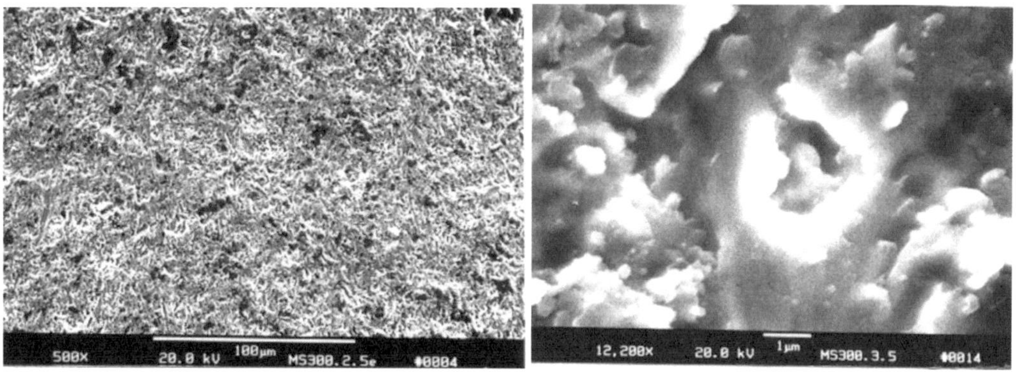

(a) (b)

Fig. 7. Scanning electron micrograph of mild steel exposed in the FB rig at 300^{0}C and 2.5 m s^{-1} showing a) the surface morphology and b) morphology at higher magnification

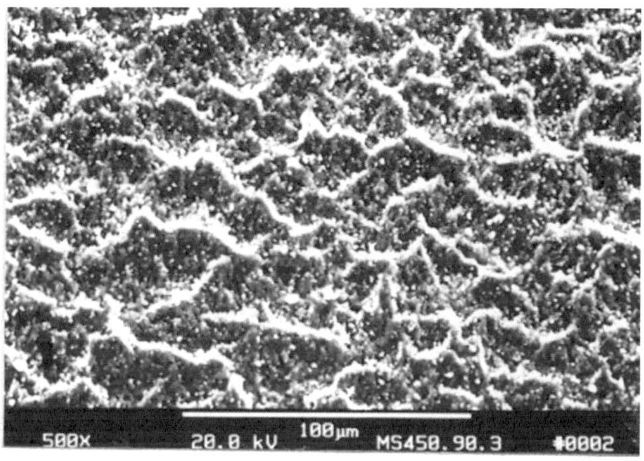

Fig. 8. Scanning electron micrograph of mild steel exposed in the FB rig at 450^{0}C and 2.5 m s^{-1}.

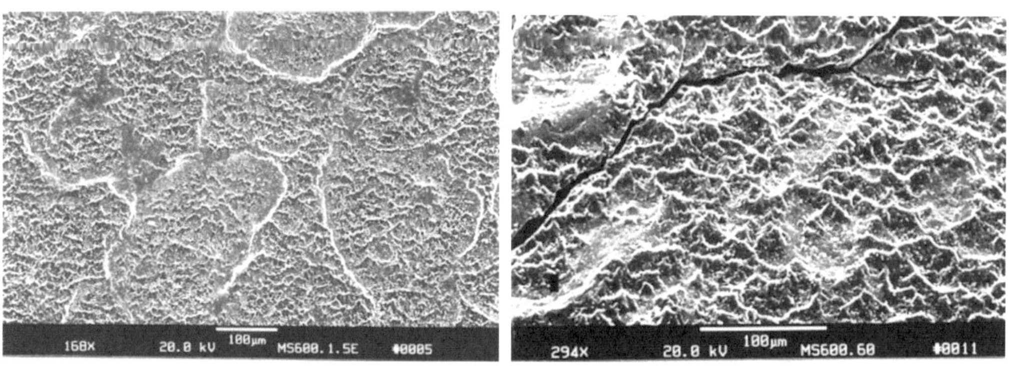

(a) (b)

(c)

Fig. 9. Scanning electron micrograph of mild steel exposed in the FB rig at 600ºC and at a) 1.5 m s^{-1}, b) 2.5 m s^{-1}, and c) 4.5 m s^{-1}

3.1.2 310 stainless steel

A most distinctive feature of the weight loss of this steel as a function of temperature when compared with the mild steel is that, no weight loss peak was recorded, under any condition, Fig. 10. This is in contrast with previous laboratory studies from other groups in the early 1990s (Ninham et al., 1990, Stack et al., 1991, Stott et al., 1990) and this is attributed to the lower impact energies of these studies. The onset of weight loss over all the temperature range was found to be strongly dependent on impact velocity. For example, at 1 m s^{-1} no weight loss was recorded at temperatures below 500ºC. Increasing velocity reduced quite significantly this "threshold" temperature. At 2 m s^{-1} it fell to 300ºC, and at 4.5 m s^{-1} it was less than 100ºC. Above this "threshold" temperature, the weight loss increased non-linearly with temperature in all cases. At 100ºC, the behaviour was similar to that of the mild steel, i. e., below 2.5 m s^{-1} no weight loss was observed, and, above this velocity, the weight loss increased non-linearly. For this steel, the extent of weight loss increased with increasing temperature for a given impact velocity. It is interesting to note that, at the lower velocities (1-1.5 m s^{-1}) the weight changes were relatively independent of velocity for both steels. However, at the high velocities (i.e. >1.5 m s^{-1}) the increase in wastage rate for mild steel as a function of increasing velocity was much greater than for the 310 stainless steel. Weight gains were recorded at the lower velocities for both materials. Indeed, this was the case for 310 stainless steel even at 2.5 m s^{-1}. At 600°C, the increase in wastage rate as a function of velocity differed from that at the lower temperature, while the overall rates were much higher than at 300°C. In the lower velocity range (1-2 m s^{-1}) the erosion-corrosion rate of the 310 stainless steel was higher than that of the mild steel. However, the results also showed that the ranking order of degradation rates of the alloys changed as a function of velocity. For example, at 2 m s^{-1} the weight loss of 310 stainless steel was approximately a factor of 10 greater than for the mild steel. At 2.4 m s^{-1}, however, there was no difference between the wastage rates of both materials, while above this velocity, the relative wastage rates of the alloys reversed, with the mild steel now giving a higher value than the 310 stainless steel.

In general, impact velocities up to and below 2.5 m s^{-1} produced rippled surfaces at all temperatures. Figures 11(a), 11(b) and 11(c) show examples of surface morphologies after exposure at 300ºC and 1.5 m s^{-1}, 450ºC and 2.5 m s^{-1}, and 600ºC and 1.5 m s^{-1}, respectively.

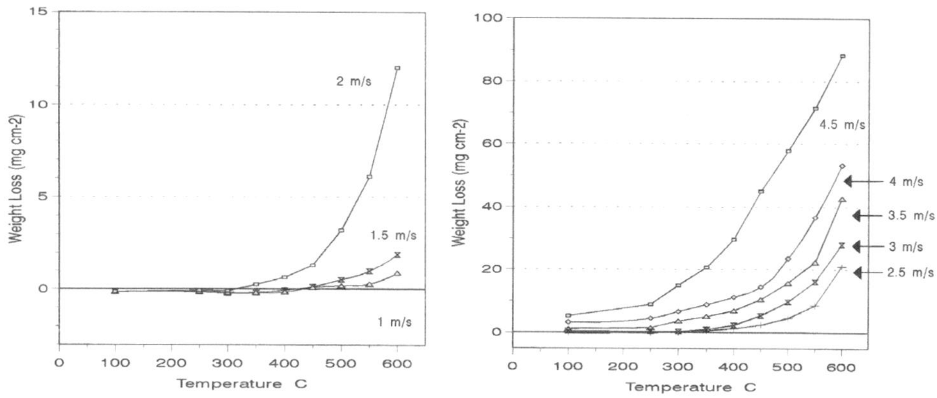

Fig. 10. Weight change as a function of temperature and specimen velocity for 310 SS exposed in the FB rig with the 560 μm alumina particles for 24 h.

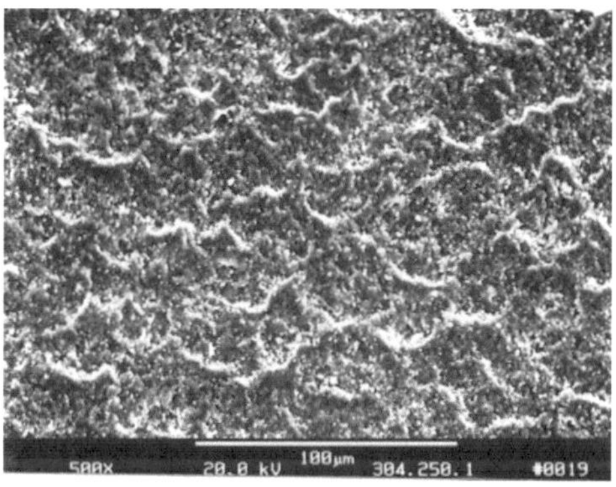

(a)

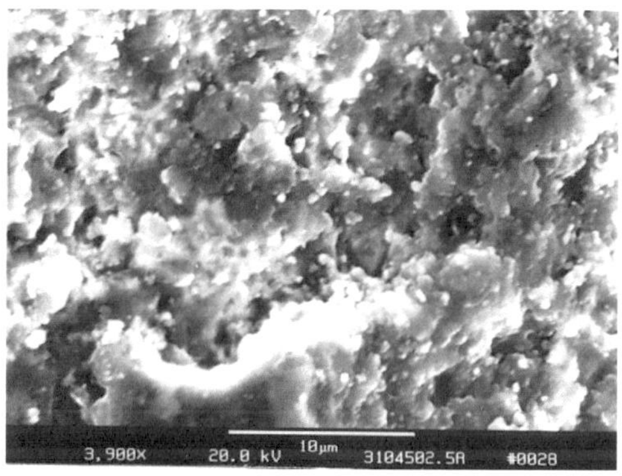

(b)

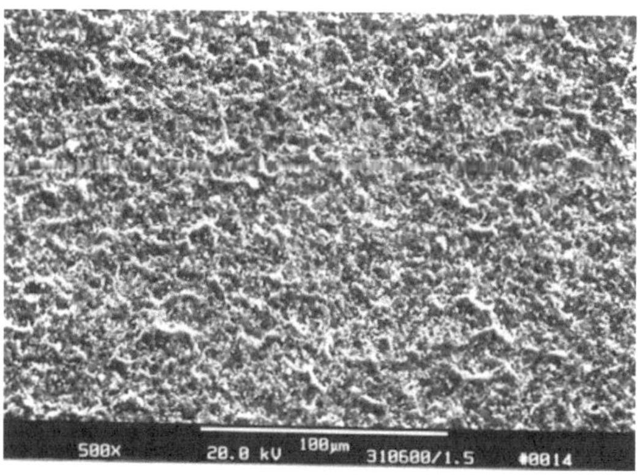

(c)

Fig. 11. Scanning electron micrograph of 310 stainless steel exposed in the FB rig at a) 300^{0}C and 1.5 m s^{-1}, b) 450^{0}C and 2.5 m s^{-1}, and c) 600^{0}C and 1.5 m s^{-1},

However, such ripples were not observed at impact velocities above 2.5 m s^{-1}. As an example of this, Fig. 12 (a and b) shows the surface morphology after exposure at 600^{0}C and 4.5 m s^{-1}, where wear tracks of about 10 to 20 μm in length and areas of scale spallation could be observed.

(a)

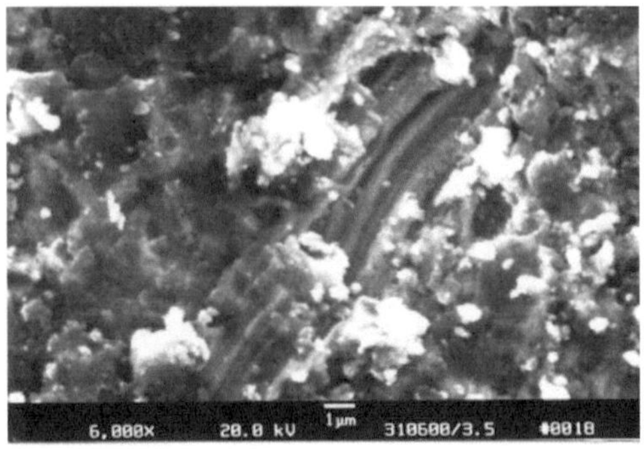

(b)

Fig. 12. Scanning electron micrograph of 310 stainless steel exposed in the FB rig at a) 600^{0}C and 4.5 m s^{-1}, b) magnification of a)

3.2 Wear pattern

Fig. 13 shows the angular dependence on thickness loss for mild steel and 310 SS at 100^{0}C. In the first case, two wear profiles are shown: at 1.5 m s^{-1}, a negative thickness was recorded in the angular range from 0^0 to 180^0. More likely, this pattern could be associated with deposition of erodent on the exposed area, as was confirmed by EDX analysis on the specimen surface. Indeed, this was the case for both steels at such low velocity and below 300^{0}C. Increasing the impact velocity produced a typical M pattern with two peaks located at each side from the front of the specimen. For instance, at 4.5 m s^{-1}, the maximum thickness loss was about 30 μm, although some material loss was also evident at normal impact angle. For the 310 SS, a similar trend was found, although the thickness loss was slightly less at both shallow and normal impact angles, compared with the mild steel. For both steels, the angle of maximum attack was almost the same, about 35^0 (145^0). The angle in the parenthesis corresponds to the second peak of the pattern.

For mild steel at 300^{0}C and 450^{0}C, and impact velocities from 2.5 m s^{-1} to 4.5 m s^{-1}, the wear patterns found are shown in Fig. 14. At 300^{0}C, it can be seen that the maximum angle of wear shifts slightly to lower angles i.e., from 33^0 to 27^0 (147^0 to 153^0) with increasing speed. This also caused a large increase in thickness loss. At the highest speed, a maximum loss of about 115 μm was recorded, whereas the specimen front has a typical loss of 15 μm (a loss was not observed at lower speeds). At 450^{0}C, and 2.5 m s^{-1}, no thickness loss was recorded, but further increases in impact velocity produced similar profiles to the ones observed at 300^{0}C. At the highest velocity, the wastage at the specimen front was somewhat higher, with a V shape pattern. At shallow angles, the main difference at both temperatures was the thickness loss magnitudes. Fig. 15 shows the results at 600^{0}C for mild steel. Here the angle of attack shifted from 33^0 to 24^0 (147^0 to 156^0) on increasing the impact velocity from 2.5 m s^{-1} to 4.5 m s^{-1}.

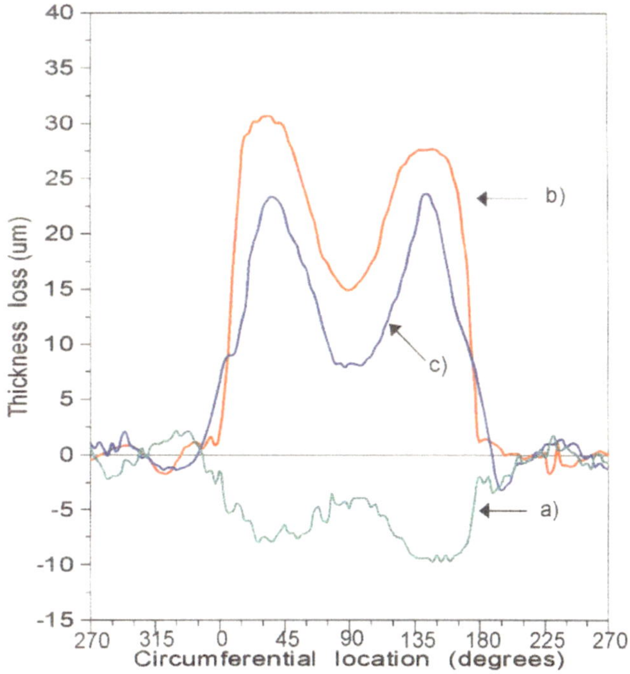

Fig. 13. Thickness loss as a function of impact angle for mild steel at 100^{0}C eroded with the 560 μm alumina particles at a) 1.5 m s^{-1}, b) 4.5 m s^{-1} and c) 310 SS at 4.5 m s^{-1}

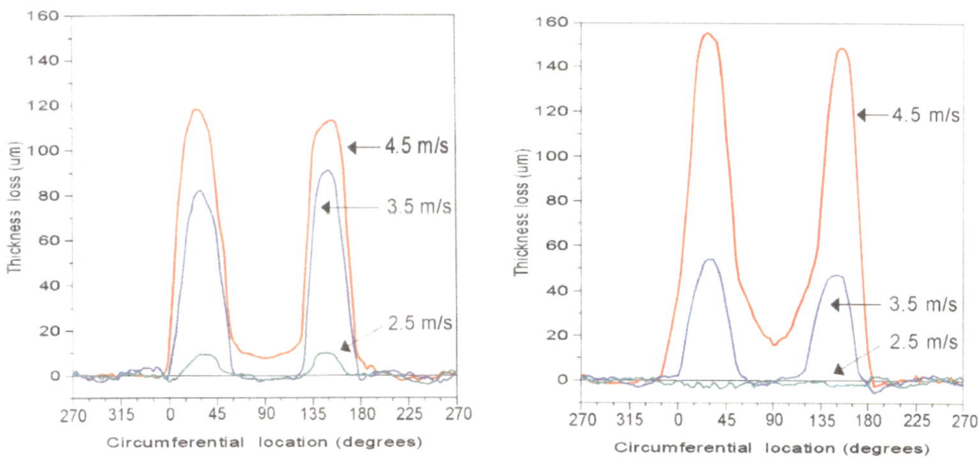

(a) 300^{0}C (b) 450^{0}C,

Fig. 14. Thickness loss as a function of impact angle for mild steel at a) 300^{0}C and b) 450^{0}C, eroded with the 560 μm alumina particles.

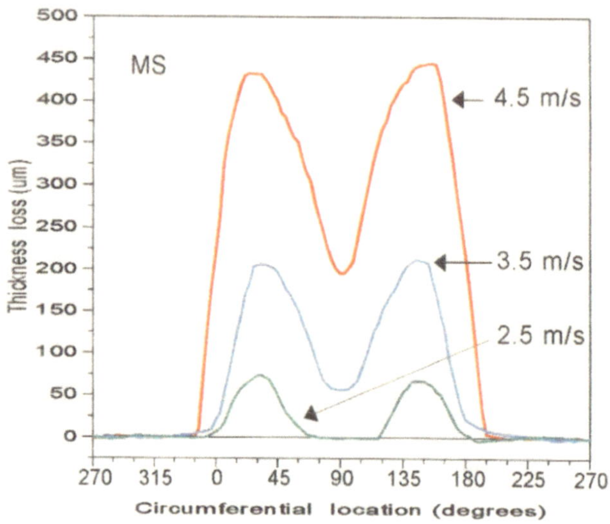

Fig. 15. Thickness loss as a function of impact angle for mild steel at 600⁰C eroded with the 560 μm alumina particles.

The wear profiles for the 310 SS as function of impact velocity at 300⁰C and 450⁰C are shown in Fig. 16. In general, on increasing velocity from 2.5 m s^{-1} to 4.5 m s^{-1}, the angle of attack changed from 33⁰ to 27⁰ (147⁰ to 153⁰) at 300⁰C, and from 30⁰ to 24⁰ (147⁰ to 156⁰) at 450⁰C. The patterns were quite similar, but it can be noted that, for each velocity, the wastage increases with increasing temperature. At 600⁰C the maximum angles of attack were about the same as the ones at 450⁰C, although there was a significant increase in the angular range of attack at both sides of the front, in particular at the highest speeds, Fig. 17.

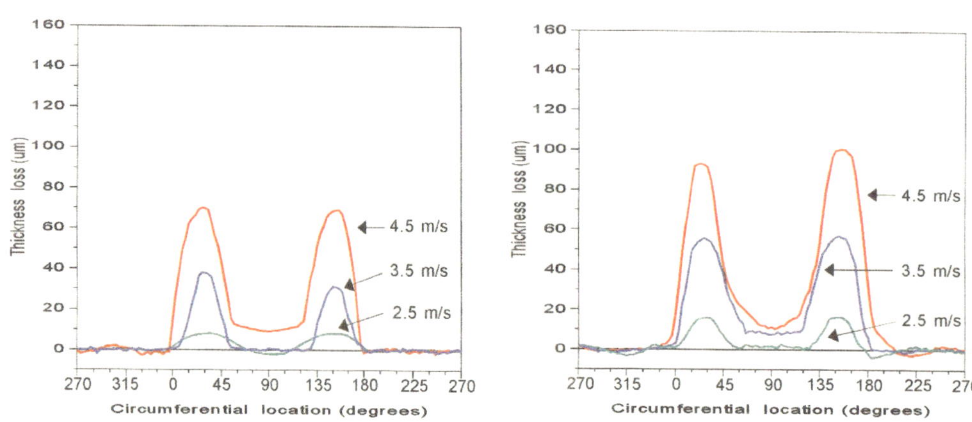

(a) 300⁰C (b) 450⁰C,

Fig. 16. Thickness loss as a function of impact angle for 310 SS at a) 300⁰C and b) 450⁰C, eroded with the 560 μm alumina particles.

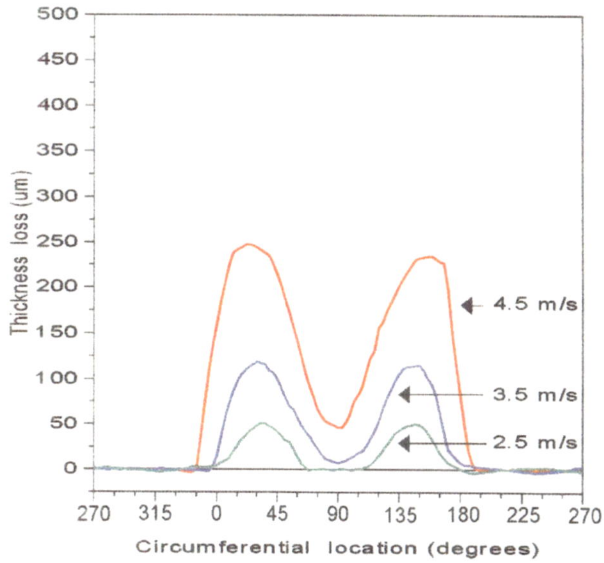

Fig. 17. Thickness loss as a function of impact angle for 310 SS at 600^{0}C eroded with the 560 μm alumina particles.

Regarding the location of maximum wastage observed in FBCs, this typically occurs at the tube bottom (90^0, Class B), or at about 60^0- 70^0 (Class A) on either side of the tube bottom, as depicted in Figure 3. In the present study, the angle of maximum wear was in the range from 20^0- 35^0. These values are small compared with the previous A classification, but this could be due to the possible differences in particle flow. On the other hand, the present results on the maximum angle of wear are in good agreement with the wear pattern reported on evaporator tubes in FBC units (Parkinson et al., 1985, Tsutsumi et al., 1989).

3.3 Effect of impact angle and mode of wear

Due to the nature of the bed, the environment in a fluidized bed may produce wear which can resemble either a three-body abrasion process or an erosion process. In the former, particles can be pressed against each other and slide over the tube surface, whereas, in the latter, particles act independently of each other, leaving the surface after impact. Although some valuable information on bed behavior has been reported in the last few years, there is still controversy about which process may dominate. On the basis of the results obtained in section 3.2, an attempt is made to describe the bed environment in terms of firstly, a dominant abrasion process, and, secondly, an erosion process. In the first case, according to Rabinowicz (Rabinowicz, 1965) the abrasive wear process is considered to be proportional to the contact pressure exerted by the particle flow multiplied by the local velocity on the specimen surface, whereas, in the second, Finnie's erosion theory has been considered (Finnie, 1960, Finnie, 1972).

3.3.1 Bed environment in the FB rig

Since the FB rig was operated at 1.3 x U_{mf}, gas velocities of about 0.17 m s^{-1} were achieved when using the larger particles. Under these conditions, Tsutsumi's equation (eq. 1), predicts particle velocities of about 0.89 m s"1. Therefore, the velocity of particles in the bed is assumed to have no significant effect on the wear of specimens, which, in turn, is dependent only on the velocity of the specimens.

3.3.2 Abrasive wear

The flow regime in the bed is described by the Reynolds number, Re. Here, a continuous medium is considered; therefore

$$\mathrm{Re} = \frac{2\rho_b * U_0 * a}{\mu_b} \tag{3}$$

Where: ρ_b = bed density, in the present case ~ 2300 kg m^{-3}; U_0 = specimen velocity relative to the particles, m s^{-1} ; a = specimen radius = 0.003 m ; μ_b = bed viscosity ~ 1.2 kg m^{-1} s^{-1} (value for a bed with 500 μm average size silica particles, (Grace, 1970)

Considering the velocity extremes in this work, i.e. 1 m s^{-1} and 4.5 m s^{-1}, this gives Re = 11.5 and Re = 51.7, respectively. In order to estimate the contact pressure, it is assumed that the particle and gas flows are uniform and behave as continua Fig. 18 shows a schematic diagram of the system under consideration.

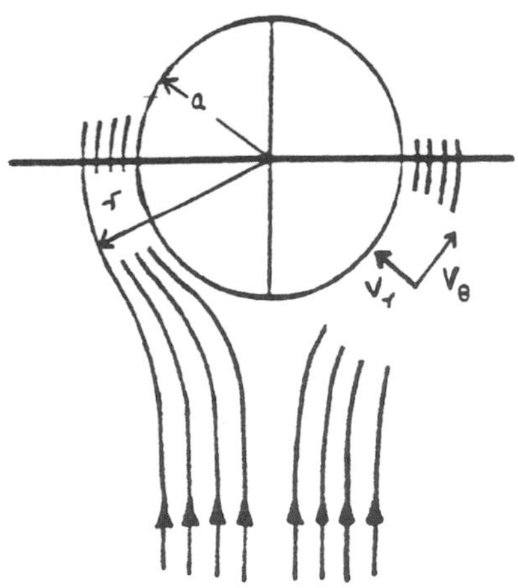

Fig. 18. Schematic diagram of flow pattern on specimens inside the bed of particles, indicating the components of velocity, Vr and V_θ.

Here, the potential fluid flow function, Φ, (in plane polar coordinates) is expressed by (Douglas et al., 1984, Kay & Nedderman, 1974)

$$\Phi = U_0\left(r + \frac{a^2}{r}\right)\cos\theta \tag{4}$$

Where r represents a point in a streamline, θ is the angle considered and a is the specimen radius. For flow around a cylinder, the radial ($V_{r\,(p,g)}$) and tangential ($V_{\theta\,(p,g)}$)components of velocity for the particles and fluidizing gas in terms of velocity potential are given by

$$V_{r(p,g)} = \frac{\partial \Phi}{\partial r} = U_0\left(1 - \frac{a^2}{r^2}\right)\cos\theta \tag{5}$$

$$V_{\theta(p.g)} = \frac{1}{r}\frac{\partial \Phi}{\partial \theta} = -U_0\left(1 + \frac{a^2}{r^2}\right)\sin\theta \tag{6}$$

At the surface of the cylinder, r = a , hence

$$V_{r(p,g)} = 0 \tag{7}$$

$$V_{\theta(p,g)} = -2U_0 \sin\theta \tag{8}$$

Now, from Bernoulli´s equation, the pressure exerted by the particle (P_p) and gas (P_g) is given respectively by:

$$P_p = P_\infty + \frac{\rho_p}{2}\left(U_0^2 - V_\theta^2\right) \tag{9}$$

$$P_g = P_\infty + \frac{\rho_g}{2}\left(U_0^2 - V_\theta^2\right) \tag{10}$$

The difference between (6.8) and (6.9) gives the net pressure acting on the cylinder surface, thus

$$P_p - P_g = \frac{1}{2}\left(\rho_p - \rho_g\right)\left(U_0^2 - V_\theta^2\right) \tag{11}$$

$$P_p - P_g = \frac{U_0^2}{2}\left(\rho_p - \rho_g\right)\left(1 - \left(\frac{V_\theta^2}{U_0^2}\right)\right) \tag{12}$$

Now, from eqn. (6.4), and assuming that, in general, the gas density is much lower than the particle density, the pressure distribution on the cylinder is given by

$$C_p = \frac{P_p - P_g}{\frac{1}{2}\rho_p U_0^2} = 1 - 4\sin^2\theta \tag{13}$$

Where Cp is the pressure coefficient. A plot of equation 13 is given in Fig. 19(a), where it can be seen that the pressure distribution is similar to that found at low Re numbers, Fig. 19(b). From classical abrasion theory (Archald, J. 1953, Rabinowicz, E. 1960), the abrasive wear rate, AWR, can be expressed as

$$AWR \propto V_\theta\left(P_p - P_g\right) \tag{14}$$

And, from eqns. (8) and (13)

$$AWR \propto U_0^3 \rho_p \sin\theta(1 - 4\sin^2\theta) \tag{15}$$

From this equation, Fig. 20 (curve A) shows the dependence of the predicted AWR on impact angle. The maximum wear is about 15° from the stagnation point, which is close to the specimen front. Hence, the above equation is not in agreement with the actual observations, where the maximum wear was observed in the range 24°to 33°. In a bubbling fluidized bed, a schematic diagram of the interaction between bubbles and a tube can be seen in Fig. 21. The region in the lower part of the bubble is the wake region, which carries entrained particles. Here, the measured wake angle is, $\theta_w \approx 120°$, as depicted in Fig. 21, for silica sand particles of 500 μm mean size (Rowe & Partridge, 1965). To some extent, when a bubble passing a tube, it appears that this angle may have some effect on the location of wear. Equation 15 thus may be modified by taking into account an angle given by $\theta_{wa} = \theta_w/2 \approx 60°$. Consider a case where the flow is modified by changing θ by θ - (θ_{wa} - 15°), where the 15° value is the difference of the angle at normal impact and the angle of maximum wear previously found. The result can be seen in fig. 20 (curve B), where the maximum angle of wear appears at about 60°, which corresponds with a 30° angle for the actual specimens, and correlates well with the experimental findings. However, it predicts higher wear at normal impact angle, which is not the case for the specimens in the present work. Now, suppose θ is changed by an amount equal to (θ - θ_{wa}) in equation 15. The result is given in Fig. 20 (curve C). Here, the maximum angle of wear at shallow angles is about 15°, this being lower than the observed range. Also, this last change predicts even a higher wear rate at the front of the specimen.

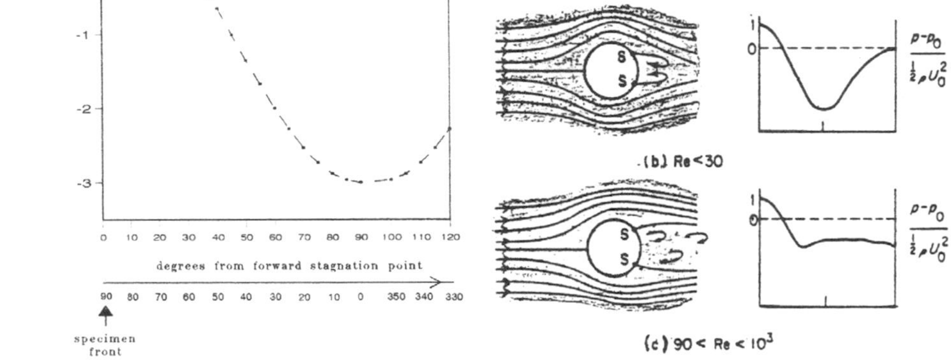

(a) (b)

Fig. 19. Coefficient of pressure, C_p, on the surface of a cylinder as a function of angle, a) as derived from equation 13, b) for flow past a cylinder for various Re ranges

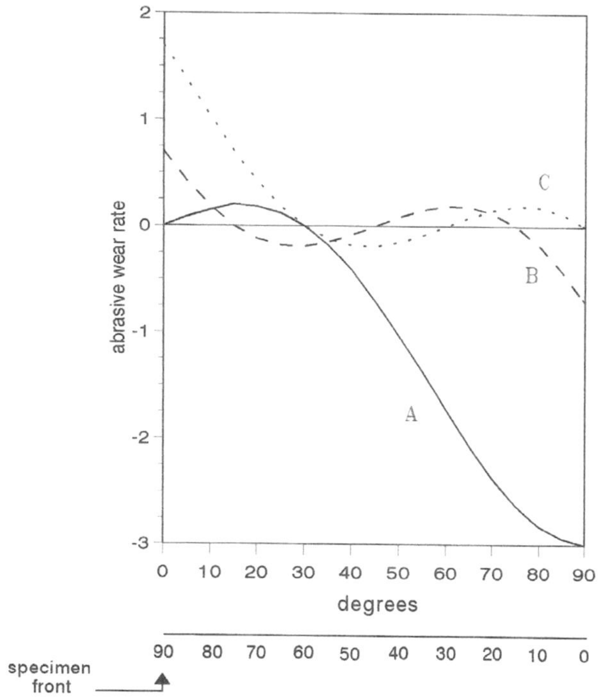

Fig. 20. Variation in abrasive wear rate as a function of impact angle, as predicted by abrasion theory: i) curve A resulting from equation 15; ii) curve B when changing θ by θ-(θ_{wa}-15°) in equation 15, and iii) curve C, changing θ by (θ-θ_{wa}) in equation 15.

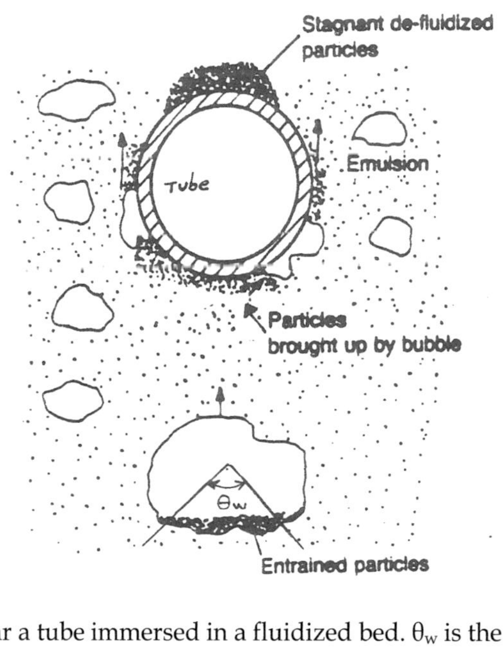

Fig. 21. Conditions near a tube immersed in a fluidized bed. θ_w is the wake angle.

3.3.3 Erosive wear

Finnie's classical erosion theory has been used in an attempt to correlate the present results on the effect of impact angle with the extent of erosion. The erosion rate expressed as volume loss, E_v , is related to the particle impact velocity, U, and angle of impingement, θ, by

$$E_v \propto U^n \cos^2 \theta \tag{16}$$

Where n is the velocity exponent (usually between 2 and 3). However, when the erosion rate is estimated by thickness loss, E_t , equation (16) may simply be multiplied by sin θ (Finnie, I. 1960). Since the results for the effect of impact angle were given as thickness loss, the resulting expression is

$$E_t \propto U^n \cos^2 \theta \sin \theta \tag{17}$$

This equation is plotted in Fig. 22. As would be expected, the predicted peak erosion angle, θ_{max} is ≈ 35°, which is in good agreement with the experimental results. However, it is worth noting that these results, Figs. 13 to 17, showed that, at low temperature, i.e. 100°C, and 4.5 m s^{-1}, θ_{max} was found at about 35°, whereas at 2.5 m s^{-1} and 300°C, θ_{max} was very similar to the previous value. This suggests that, at low temperatures, the impact velocity apparently had little effect on the peak erosion angle. At higher temperatures, θ_{max} shifted slightly to lower angles with increasing velocity from 2.5 m s^{-1} to 4.5 m s^{-1}. This is because changes in particle flow may take place as a function of both temperature and velocity. At the highest velocity used, increasing temperatures also shift θ_{max} to lower angles, (see for example Figs. 13 and 15). This suggests that the steels exhibit a more ductile behavior with temperature. Another observation is that, at least for the alloys studied, θ_{max} seems to be independent of the steel type.

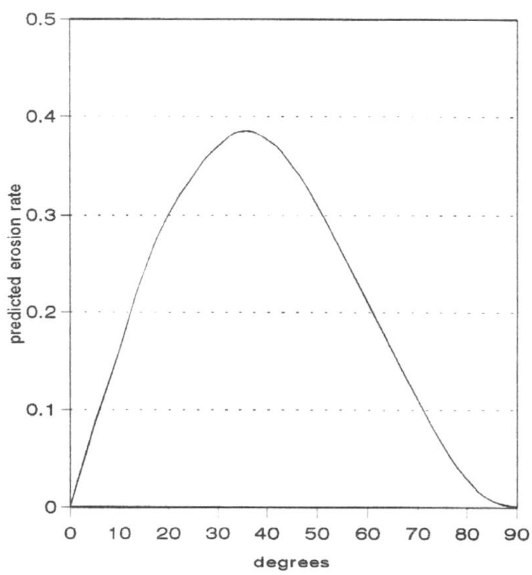

Fig. 22. Predicted erosion rate as a function of impact angle according with Finnie´s erosion theory. The angle of maximum attack is about 35^0. The specimen front is at the 90^0 angle.

In general, with the larger particles, ripple formation was a typical feature (which is generally associated with a purely erosion process) for most steels exposed at velocities up to about 3 m s^{-1}. Higher velocities produced surface morphologies that were dependent on the type of steel. For instance, for the low alloy steels, polished surfaces were not uncommon, whereas, for the stainless steels, a more clear ploughing and cutting action was observed. The wear tracks developed mainly on the 310 SS are consistent with a dominant abrasion process. However, very similar morphologies (showing cutting/ploughing action) for low alloy steels and stainless steels have been reported in tests carried out in more conventional rigs (Morrison et al., 1986, Zhou & Bahadur, 1990).

3.3.4 Comparison between abrasion and erosion theories

Based on the modifications made in the abrasion theory, leading to the behaviour given by curve B, Fig. 20, both theories predicted about the same peak angle wear, but only at low impact angle. The abrasion theory predicts higher wastage rates at normal impact angle, i. e. at the specimen front, while the erosion theory predicts no wastage at this location. This was normally the case at velocities typically bellow 4.5 m s^{-1}. Under mild fluidization conditions in a cold FB rig, it was found the formation of an air film directly below a cylindrical obstacle (Glass & Harrison, 1964). At this location (stagnant point) the gas and particle flows are at minimum values. Another possibility is the preferential embedment and deposition of very fine erodent particles that occur at normal impact angle compared with low angles. This may have modified the wastage rate here due to a shield effect. However, in general, exposure at temperatures above 300^{0}C and velocities above 2.5 m s^{-1} produced some wastage at the front of the specimens, clearly minimizing the previous effects. Taking the angular distribution of wear as a reference, it is very difficult to determine which form of wear may dominate, since a FB environment includes a dilute phase erosive condition and also a dense (continuous) phase abrasive condition. Disagreement between researchers is not uncommon; for instance, in one study, it was suggested that abrasion is responsible for the wear at the bottom of tubes (Stringer & Wright, 1987); however, at the same location another report concluded that wastage is by erosion only (Wheeldon, 1990). On the basis of purely morphological features, the present results suggest that the main form of wear is one of an erosive nature. This agrees well with the results obtained in a FB rig facility, where erosion was the main form of wastage, but with a small amount of three-body abrasion contributing to the damage (Wang et al., 1993)

4. Conclusions

1. A temperature of peak wastage (PWT) was observed for mild steel at about 300^{0}C but only within a certain velocity range i e., 2.5 m s^{-1} < PWT ≤ 4 m s^{-1}. The wastage of the 310 SS as a function of temperature did not show any peak wastage for the velocities studied in this work.
2. In general, erodent deposition was a dominant process at impact velocities below 3 m s^{-1} and temperatures below about 300^{0}C, regardless of the type of steel. The impact angle at which wear was a maximum was about 20^0-30^0 on each side of the leading point. Regarding this observation, it is worth noting that the results showed that at the lower temperature and the highest velocity used, the peak wastage angle, θ_{max}, was found at about 35^0, whereas at 300^{0}C and 2.5 ms^{-1} was very similar. At low temperatures, the

impact velocity apparently had little effect on the peak wastage angle. At higher temperatures, θ_{max} shifted slightly to lower angles with increasing velocity from 2.5 ms^{-1} to 4.5 ms^{-1}. At the highest impact velocity used, increasing temperatures also shift θ_{max} to lower values. This suggests that the steels exhibit a more ductile behavior with temperature.
3. A further observation is that, at least for the steels used here, θ_{max} seems to be independent of the steel type.
4. On the basis of wear patterns found as a function of impact angle, an attempt has been made to define the probable modes of wear i.e. abrasion vs. erosion. The modified abrasion theory predicts well the wear pattern at shallow angles, but predicts higher wastage rates at normal impact angle, i.e. at the specimen front. On the other hand, erosion theory predicts maximum wear at an impact angle of 35^0 and no wear at the specimen front. Thus both theories have drawbacks with respect to damage in the FB rig.
5. Under the conditions of temperature and velocity considered, the wear losses were greater at shallow impact angle compared with the specimen front diminishing the importance of abrasion.
6. Following exposure to the test conditions, the formation of ripples, which are a feature of a purely erosive process, were often observed. Wear tracks were also observed, in particular at the highest velocities. These last features could be related to abrasive wear.

5. Acknowledgments

The authors are grateful to CONACyT (Mexico) for supporting this work. Also we would like to express our thanks to Dr. Jiang Jiaren for permission to use his profilometer system, and to Gabriela K. Pedraza-Basulto, MSc., Adan Borunda-Terrazas, MSc., and Gregorio Vazquez-Olvera, MSc., for technical support.

6. References

Anderson, J., Carls, E. Mainhardt, P., Swift, W. Wheeldon, J., Brooks, S. Minchener, A. & Stringer J. (1987).Wastage of In-Bed Heat Transfer Surfaces in the Pressurized Fluidized Bed Combustor at Grimethorpe. *J. Eng. Gas Turbines and Power,* Vol.109, No.3, (July 1987), pp. 298-303, ISSN 0022-0825

Archald, J. (1953). Contact and Rubbing of Flat Surfaces. *J. Appl, Physics,* Vol.24, No.8, (Aug. 1953), pp. 981-988, ISSN 0021-8979

Beacham, B. & Marshall, A. (1979). Experiences and Results of Fluidized Bed Combustion Plant at Renfrew. *J. Inst. Energy,* Vol.52, No.411, (July 1979), pp. 59-64, ISSN 0144-2600

Boiarski, A. (1978) Testing, Identification and Evaluation of Advanced Experimental Materials and Coatings in the Design Conditions for Simulated Fossil Fuel Power Cycle Conditions, Task 2, *Batelle Report NTIS No. FE-2325-19, DOE Contract E(49-18)-2325, Dec 1978*

Douglas, J., Gasiorek, J. & Swaffield, J. (1984). Fluid Mechanics, 2nd Edition, Pitman Inst. Texts, London, ISBN-10 0273021346

Finnie, I. (1960). Erosion of Surfaces by Solid Particles. *Wear,* Vol.3, No.2, (March 1960), pp. 87-103, ISSN 0043-1648

Finnie, I. (1972). Some Observations on the Erosion of Ductile Metals. *Wear,* Vol.19, No.1, (Jan 1972), pp. 81-90, ISSN 0043-1648

Glass, D. & Harrison, D. (1964). Flow Patterns Near a Solid Obstacle in a Fluidized Bed. *Chem. Eng. Sci.,* Vol.19, No.12, (Dec. 1964), pp. 1001-1002, ISSN 0009-2509

Grace, J. (1970). The Viscosity of Fluidized Beds. Can. J. Chem. Eng., Vol.48, No.1, (Jan 1986), pp. 30-33, ISSN 1939-019X

Highley, J. & Kaye, W. (1983). Fluidized Beds Industrial Boilers and Furnaces, *Fluidized Beds: Combustion and Applications,* J.R. Howard (Ed.) 77-169, ISBN-0-8247-4699-6

Hou, P., MacAdam, S., Niu, Y. & Stringer, J. (2004). High Temperature Degradation by Erosion-Corrosion in Bubbling Fluidized Bed Combustors. *Materials Research,* Vol.7, No.1, (March 2004), pp. 71-80, ISSN 1516-1439

Howard, J.R., (1989) *Fluidization Technology,* Adam, Hulger, Bristol and New York. ISBN 0-85274-055-7

Janson, S. (1985). *Tube Wastage Mechanisms in Fluidized Bed Combustion Systems. Proceedings 8th Int. Conf. On Fluidized Bed Combustion*, pp. 750-759, ISSN: 0197453X, Springfield, Va, USA, March 18-21, 1985

Kantesaria, J. & P. Jukkola, D. (1983) Observations on Erosion of In-Bed Tubes in the Great Lakes AFBC. *Materials and Componenst Newlwtter US DoE,* No.46, (July 1987), pp. 6

Kay, J. & Nedderman, R. (1984). Fluid Mechanics and Heat Transfer, 3rd Ed. Cambridge University Press, ISBN-0521303036

Levy, A. (1987). , The Erosion-Corrosion of FBC, In–Bed Tubing Alloys, *EPRI Worshop on Wastage on In-Bed Surfaces in Fluidized Bed Combustors, Vol III* , pp. 1-73, Argonne, Illinois, USA, Nov 2-6, 1987

Mezko, J. (1977). *Metal Progress,* Vol. 112, pp.30

Montrone, E. (1987). Experience with Foster Wheeler FBC´S, *EPRI Worshop on Wastage on In-Bed Surfaces in Fluidized Bed Combustors, Vol III* , pp. 1-11, Argonne, Illinois, USA, Nov 2-6, 1987

Morrison, C., Scattergood, R. & Routbout, J. (1986). Erosion of 304 Stainless Steel. *Wear,* Vol.111, No.1, (Jan 1986), pp. 1-13, ISSN 0043-1648

Ninham, A. J., Hutchings, I. M. & Little, J. A. (1990). Erosion-Oxidation of Austenitic and Ferritic Alloys. *Corrosion,* Vol.46, No.4, (Apr 1990), pp. 296-301, ISSN 0010-9312

Oka, S. (2004). Fuidized Bed Combustion, Marcel Dekker, New York. ISBN 0-8247-4699-6

Parkinson, M., Napier, B., Jury, A. & Kempton, T. (1985). Cold Models Studies on PFBC Erosion, *Proceedings of 8th International Conference on Fluidized Bed Combustion Vol II,* pp. 730-738, ISSN: 0197453X, Springfield, Va, USA, Mach 18-21, 1985

Peeler, J. & Whitehead, A. (1982). Solids motion at horizontal tube surfaces in a large gas-solid fluidized bed *Chem. Eng. Sci.,* Vol.37, No.1, (Jan 1982), pp.77-82, ISSN 0009-2509

Rabinowicz, E. (1995). Friction and Wear of Materials, John Wiley & Sons, ISBN 0471830844

Rademarkers, P. & Kettunen, P. (1996). Materials Requirements and Selection for Fluidized Bed Combustors , *High Temperature Alloys for Gas Turbines and Other Applications 1986, Part 2,* W. Muntz (Ed.) 269-292, ISBN-13 978-9027723482

Rademarkers, P., Lloyd, D. & Regis, V. (1990). ABFC´S: Bubbling, Circulating and Shallow Beds In: *High Temperature Materials in Power Energy, Part 1,* E. Bachelet, (Ed.), 43-63, ISBN 978-0792309277

Rowe, P., & Partridge, B. (1965). An X-Ray Study of Bubbles in Fluidised Beds. *Trans. Inst. Chem. Eng.*, Vol.43, pp. 157-175, ISSN 0043-1648

Stack, M. M., Stott, F. H. & Wood, G. C. (1991). Erosion-Corrosion of Preoxidized Incoloy 800H in Fluidized Bed Environments: Effects of Temperature, Velocity and Exposure Time. *Mat. Sci and Tech,* Vol.7, pp. 1128-1137, ISSN 0267-0836

Stott, F. H., Stack, M. M. & Wood, G. C. (1990). The Role of Oxides in the Erosion-Corrosion of Alloys Under Low Velocity Conditions, *Proceedings: Corrosion-Erosion-Wear of Materials at Elevated Temperatures,* pp. 12-1 to -12-16, ISBN 1-877914-18-5, Berkeley, Calif, USA, Jan 31-Feb 02, 1990

Stringer, J. & Wright, I. (1987). Erosion/Corrosion in FBC Boilers, *EPRI Worshop on Wastage on In-Bed Surfaces in Fluidized Bed Combustors, Vol I* , pp. 1-30, Argonne, Illinois, USA, Nov 2-6, 1987

Tsutsumi, K., Tatebayashi, J., Hasegawa, K., Takamori, M., Okada, Y. & Furobayashi, K. (1989). Development of Erosion Resistant In-Bed Tubes. *Proceedings Int. Conf. On Fluidization (Fluidization VI),* ISBN 0816904596, Banf, Alberta, Can. May 1989

Wang, B., Geng, G. & Levy, A, (1993). Effect of Microstructure on Materials Wastage in a Room-Temperature Fluidized-Bed Wear-Test Rig . *Wear,* Vol.165, No.1, (May 1993), pp. 25-33, ISSN 0043-1648

Wang, B. & Levy, A, (1992). Erosion-Corrosion of 1018 Steel at Eroded Low Velocities by Bed Material. *Wear,* Vol.155, No.1, (May 1992), pp. 137-147, ISSN 0043-1648

Wheeldon, J. (1990). A Re-Evaluation of Tube Wastage Data Collected from a Bubbling Fluidised Bed Cold Model *Proceedings: Corrosion-Erosion-Wear of Materials at Elevated Temperatures,* pp. 41-1 to 41-13, ISBN 1-877914-18-5, Berkeley, Calif, USA, Jan 31-Feb 02, 1990

Zhang, X. (1980). *Proceedings 6th Int. Conf. on Fluidized Bed Combustion,* OCLC: 6896711, Atlanta, Ga, USA, April 9-11, 1980

Zhou, J. & Bahadur, S. (1990). Futher Investigations on the Elevated Temperature Erosion-Corrosion of Stainless Steels, *Proceedings: Corrosion-Erosion-Wear of Materials at Elevated Temperatures 1990,* A. Levy (Ed.) pp 13-1 to 13-17, ISBN 1-877914-18-5, Berkeley, Calif, USA, Jan 31-Feb 02, 1990

6

Heat and Thermochemical Treatment of Structural and Tool Steels

Jan Suchánek
Czech Technical University in Prague, Czech Republic

1. Introduction

Degradation processes are apparent on the working surfaces of stressed parts, which gradually impair the parameters of machine parts and equipment. These degradation processes include abrasive wear caused by the interaction of hard, usually mineral particles with the working surface of the part. In marginal cases intensive abrasive wear leads to great losses due to output stopages caused by machine or equipment outages and further due to costs necessary for production of spare parts and/or their renovation and maintenance costs. It is often stated that approximately 50% cases of wear are due to abrasive processes (Krushchov & Babichev,1970; Eyre,1979; Uetz,1986).

Abrasive wear is characterized by separation and displacement of material particles during scratching and cutting caused by hard particles. These particles can either be free or in a certain sense bonded.

Another case of abrasive wear is known from experience in which hard particles are present between two working surfaces moving relatively one to another. This case is typical for crushing and grinding matter. However it can also be found in sliding pairs where hard impurities penetrate into insufficiently sealed working surfaces. Hard particles can also be formed in the process of adhesive wear which can affect one or both surfaces abrasively. Such a classification of types of abrasive wear was generally accepted although it appears that the processes of interaction of abrasive particles and the material subject to wear are much more complicated. What must be taken into account is the character and time response of forces acting between the abrasive particles and the surface subject to wear.

Ideally, the materials for resistance to abrasion should have the hardness that is in excess of that of the mating surface or the abrasive particles. But in the most cases the hardness of abrasive particles is higher than the hardness of abraded materials.

2. Effect of structural factors on abrasion resistance

2.1 Effect of hard structural compoments on failure of metallic materials by abrasion

The basic concept of the effect of structural factors on abrasion resistance was worked out by (Krushchov & Babichev, 1960). He raised an assumption that the mechanism of wear is identical for all structural components. The abrasion resistance of heterogeneous materials is given by the sum of products of the volume shares of individual structural phases and their relative abrasion resistance (additivity law). The thermal affection of the process by wear and by structural modifications is not considered. According to (Zum Gahr, 1985) the behaviour of heterogeneous materials is not controlled by a single phase but that the contribution of each phase is linearly proportional to its volume share. Abrasive wear decreases linearly with the growing volume of the hard phase. Both models assume that all components of the composite are subject to wear in the same way and that the contribution of each component depends only on its volume share and proportional wear. The effect of further important factors such as the properties of interphase boundaries, relative size and fracture toughness of the phases are not considered by these models although is obvious that they have a significant effect on the abrasive wear of a heterogenous material (Simm & Freti, 1989).

Abrasion resistance of metallic materials depends on the hardness, shape, size and amount of hard structural components and their distribution in the basic metallic matrix. The growing hardness of these phases and their amount in the microstructure increase the abrasion resistance of the material (Roberts et al., 1962; Schabuyev et al., 1972; Suchánek et al., 2007). The share of carbides, however, cannot by considered separately from the basic material. E.g. in pearlite-carbide microstuctures the hardness and abrasion resistance grow with the growth of the carbide volume up to 35% (Popov & Nagorny, 1970).

However in ferrite-carbide microstructures abrasion resistance grows with the share of carbides up to 5-6% and further growth does not affect abrasion resistance any more (Popov & Nagorny, 1969). Besides the share of carbides important factors are the type of the carbide phase and its size. According to (Grinberg et al., 1971), who studied abrasion resistance in steels with a constant content of Cr, W and V with a ferritic matrix and uniformly distributed carbides, the abrasion resistance of steels grows with the change of a type of carbide to a type richer in the alloy element. E.g. when changing from a carbide of the M_3C type to a more complex carbide type M_7C_3, abrasion resistance grows. Special carbides of alloy elements increased abrasion resistance compared with steels containing complex carbides (Popov et al., 1969). A typical example is the addition of vanadium in ledeburitic chromium steels which results in an increase of abrasion resistance due to formation of a very hard VC carbide (2100 - 2800 HV). The effect of the type of carbides on the relative abrasion resistance Ψ_a of quenched steels with a hardness in the range 850 - 900 HV is given in the diagram in Fig. 1 (Suchánek & Bakula, 1987). For this range of hardnesses the share of retained austenite in high-alloy materials is low and affects overall abrasion resistance only slightly. The values of Ψ_a obtained from an apparatus with an abrasive cloth showed that lowest abrasion resistance was found in steels with Fe_3C and/or M_3C carbides. Higher abrasion resistance was found in steels with M_7C_3 carbides and maximum abrasion resistance in steels containing special MC carbides.

High dispersity of carbides is most favourable for the achievement of high wear resistance (Tsypin, 1983). This is why abrasion resistance grows in tempered steels with the growing dispersity of cementite particles (Larsen-Badse & Mathew, 1966).

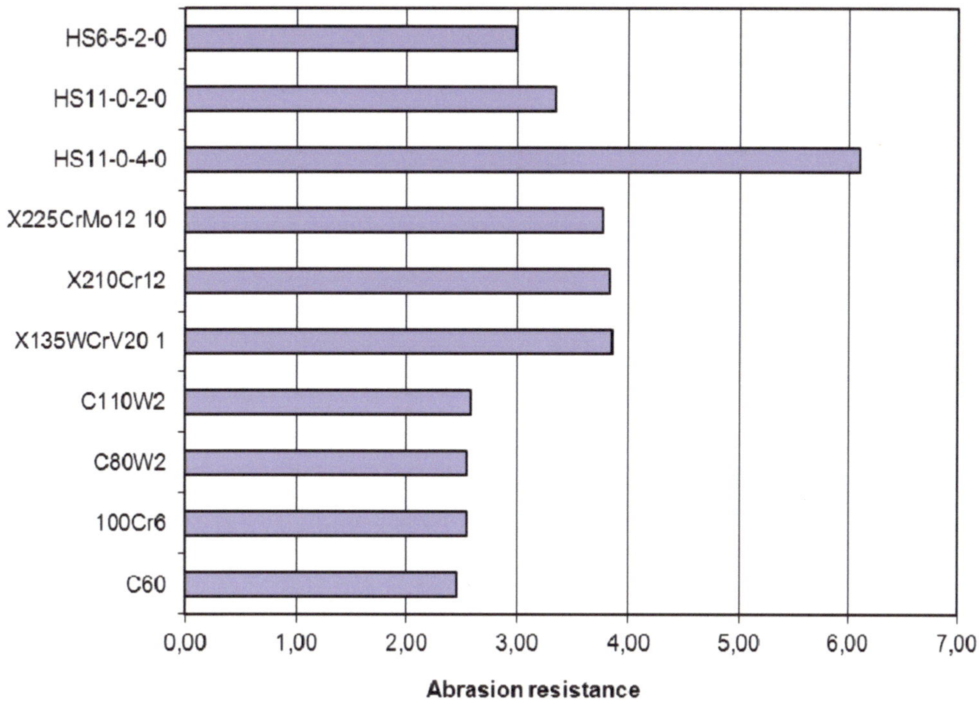

Fig. 1. Influence of the carbide types on abrasion resistance of quenched steels (100Cr6, C80W2, C110W2 - carbide M_3C; X135CrWV20 1, X210Cr12, X225CrMo12 10 - carbide M_7C_3; HS6-5-2-0, HS11-0-2-0, HS11-0-4-0 - carbides $M_6C + M_{23}C_6 + MC$) (Suchánek & Bakula, 1987).

2.2 Effect of matrix on processes of abrasion

If available references in literature are summarized, the following conclusion can be arrived at steels with a ferritic matrix exhibit the lowest abrasion resistance. Substitutional hardening of ferrite by carbide-forming elements (Cr, W, V) does not result in an enhancement of abrasion resistance. The growing share of pearlite in ferritic-pearlitic steels results in enhancement of resistance to wear. Steels and cast irons with a martensitic matrix have a higher abrasion resistance than pearlitic steels and cast irons with a similar chemical composition. The higher content of carbon and alloy elements leads to a marked enhancement of abrasion resistance (Moore, 1974; Filippova & Goldshtein, 1979).

The presence of carbides in the martensitic matrix results in a further increase of resistance to abrasive wear by particles. The hardness and quantity of these carbides makes a significant contribution to the abrasive resistance of a metal. The matrix and carbides resist penetration of abrasive particles that start a cutting action.

During heat treatment of steels and cast irons with a higher content of carbon and alloy elements, which shift M_s and namely M_f temperatures towards lower values, retained austenite is formed in the structure. A number of papers were devoted to its influence on abrasion resistance (Petrov & Grinberg, 1968; Popov & Nagorny, 1970; Fremunt et al., 1971). The authors of these papers experimentally proved in various grades of steels and cast irons that the function of retained austenite during sliding abrasion is positive. An explanation for this behaviour is that the strain-induced martensitic transformation can contribute to the enhancement of wear resistance at low impact energies.

Steel grade	Heat treatment	Microstructure	Hardness [HV]	ψ_a
RFe100	annealed	ferrite	100	1.0
C45	normalized	ferrite + pearlite	195	1.32
C80W2	soft annealed	spheroidal pearlite	167	1.19
85MnCrV8	soft annealed	spheroidal pearlite	186	1.31
X190Cr12	soft annealed	pearlite + carbides	216	1.56
X195CrVWMo5 4	soft annealed	pearlite + carbides	265	1.66
HS12-0-2-0	soft annealed	pearlite + carbides	223	1.25
C45W	quenched	martensite	789	1.96
C80W2	quenched	martensite	865	2.57
X190CrWV12	quenched	martensite + carbides	801	3.28
X195CrVWMo5 4	quenched	martensite + carbides	772	4.22
HS12-0-2-0	quenched	martensite + retained austenite + carbides	752	3.31

Table 1. Heat treatment, microstructure, hardness and abrasion resistance of selected steels ψ_a

The frequency of failures and the life of machines and equipment in operation depend on a number of factors - e.g. design, material used for significant parts, quality of production, response of the worked medium and operating conditions. In spite of the fact that basic knowledge on the mechanisms of separation of wear particles from the surface of the material subject to wear has already been obtained, the effect of heat and thermochemical treatment, microstructure of metallic materials on the intensity of wear have not as yet been fully explained. Structural effects in carbon and low alloy steels used in cases of low intensities of abrasive wear were studied in a number of published papers (Soroko-Novickaya, 1959; Krushchov & Babichev, 1960; Richardson, 1967; Khruschov, 1974; Eyre, 1979; Wirojanupatump & Shipway, 1999). Experimental work performed in this area has a considerable practical effect since it enables optimum selection of the steel and its heat and/or thermochemical treatment for parts working in conditions of abrasive wear.

3. Method of abrasive wear testing

Abrasive wear resistance of different steels was tested with the apparatus with an abrasive cloth. On the apparatus SVUM AB-1 the pin-shaped specimen with a 10 mm diameter slides against a corundum abrasive cloth. During the tests the disc with a fixed abrasive cloth (grain size 120) rotates and the specimen moves radially with a 3 mm/1 rotation.

Reference specimens (annealed plain carbon steel with 0.045% C, 95-105 HV) and tested specimens were used following a 1-2-1-2-1 scheme. Test parameters: sliding path - 50 m, specific pressure - 0,32 MPa, max. velocity - 0,53 m/s. The abrasive wear resistance ψ_a of the tested steels was estimated from the ratio of the wear volume of the reference material/the wear volume of the tested steel (Vocel & Dufek, 1976; Suchánek & Bakula, 1987).

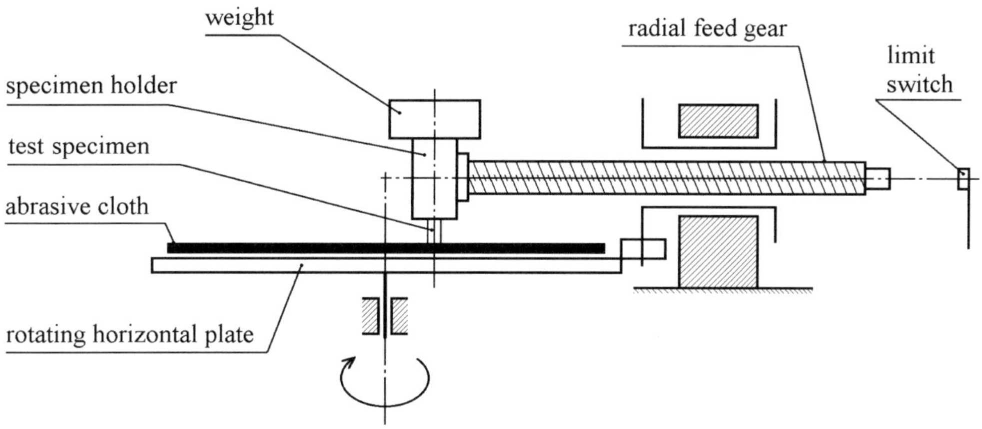

Fig. 2. Apparatus with abrasive cloth for testing abrasive wear

4. Abrasion resistance of heat treated structural steels

The effect of microstructure on abrasion resistance was experimentally tested on a selected group of currently produced structural steels. Besides heat treated carbon and low-alloyed steels, stainless steels and austenitic Mn steel (Hadfield steel) also carburizing and nitriding steels were tested. The chemical compositions of the tested structural steels are given in Table 2.

Heat treatment was applied to modify their microstructure and their physical and mechanical properties. The effect of quenching and tempering temperatures on hardness and abrasion resistance was studied in structural steels. The conditions of heat treatment can modify the microstructure and the physical and mechanical properties of low and high alloyed structural steels within a wide range. Thus consequently also their abrasion resistance changes. A summary of conditions of heat treatment, hardness and abrasion resistance of structural steels is given in Tables 3 and 4.

Grade RFe100 steel with a very low carbon content (0.045%C) used for tests of abrasive wear as a standard material has a ferritic structure after annealing.

The matrix of normalized chromium-manganese carburizing steel grade 20MnCrTi is ferritic-pearlitic. After quenching the microstructure of this steel is martensitic.

The microstructure of Cr-Ni grade 17NiCr6 4 steel after normalizing is ferritic-pearlitic. After oil quenching from 860°C the structure is martensitic. Tempering at 150°C/1h results in a small decrease in the hardness and abrasion resistance.

Steel grade	Content of elements (%)							
	C	Mn	Si	Cr	Ni	Mo	V	Others
RFe100*	0.045	0.24	0.05	--	--	--	--	--
20MnCrTi	0.20	1.08	0.34	1.10	--	--	--	0.11 Ti
17NiCr6 4	0.16	0.91	0.31	0.94	1.51	--	--	--
33CrAl6	0.36	0.82	0.21	1.58	--	--	--	1.32 Al
30CrMoV9	0.27	0.56	0.30	2.56	0.3	--	0.23	0.35 Al
37MnSi5	0.35	1.32	1.29	-	-	-	-	--
58CrV4	0.61	0.96	0.39	1.06	-	-	0.11	--
100CrMn6	0.96	1.02	0.50	1.45	-	-	-	--
X10Cr13	0.11	0.28	0.30	13.18	0.06	-	-	-
X45Cr13	0.42	0.56	0.49	14.40				
X4CrNi18-10	0.074	0.87	0.27	17.95	9.00			
X5CrNiMoTi17-12-2	0.104	0.47	0.22	16.30	10.70	1.58	-	0.57 Ti
X120Mn12	1.05	11.05	0.21	< 0.10	0.17	-	-	-

Table 2. Chemical composition of tested steels

Steel grade	Heat treatment	Hardness (HV)	Abrasion resistance ψ_a
RFe100	A 700°C/4h/furnace	95-105	1.00
20MnCrTi	A 870°C/air	214	1.28
	Q 850°C/oil	446	1.45
17NiCr6 4	A 900°C/air	261	1.31
	Q 860°C/oil	417	1.44
	Q 860°C/oil + T 150°C/1h/air	413	1.40
30CrMoV9	A 840°C/air	390	1.46
	Q 880°C/oil	495	1.53
	Q 880°C/oil + T 630°C/1h/water	341	1.36
33CrAl6	A 900°C/air	290	1.49
	Q 880°C/oil	520	1.78
	Q 880°C/oil + T 630°C/1h/oil	280	1.38
37MnSi5	A 700°C/4h/furnace	179	1.21
	A 870°C/air	225	1.29
	Q 960°C/oil	485	1.66
	Q 960°C/oil + T550°C/1h	276	1.33
58CrV4	A 720°C/4h/furnace	189	1.40
	A 870°C/air	480	1.89
	Q 820°C/water	825	2.22
	Q 860°C/oil	757	2.17
	Q 860°C/oil + T420°C/1h	527	1.55
	Q 860°C/oil + T670°C/1h	301	1.54
100CrMn6	A 750°C/4h/furnace	181	1.32
	Q 820°C/oil	847	2.59
	Q 820°C/oil + T160°C/3h	724	2.37

Table 3. Heat treatment of tested structural steels (A - annealing; Q - quenching; T - tempering)

Steel grade	Heat treatment	Hardness (HV)	Abrasion resistance ψ_a
X10Cr13	Q 940°C/oil	420	1.62
	Q 940°C/oil + T200°C/2h	420	1.65
	Q 940°C/oil + T450°C/2h	403	1.54
	Q 940°C/oil + T700°C/2h	203	1.27
X45Cr13	Q 1030°C/oil	673	2.17
	Q 1030°C/oil + T150°C/30min	624	2.30
X4CrNi18-10	Austenitizing 800°C/30min/water	166	1.55
X5CrNiMoTi17-12-2	Austenitizing 1060°C/30min/water	195	1.65
X120Mn12	Austenitizing 1070°C/water	220	1.59

Table 4. Heat treatment, hardness and abrasion resistance of stainless steels and Hadfield steel (Q - quenching; T - tempering)

The microstructure of Cr-Mo nitriding grade 30CrMoV9 steel after normalizing comprises ferrite and fine lamellar pearlite. After quenching and tempering to 630°C the microstructure is formed by highly tempered martensite. The decrease of the hardness (about 154 HV) is accompanied by the decrease of abrasion resistance from 1.53 to 1.36.

The microstructure of Cr-Mn-Al nitriding grade 33CrAl6 steel after normalizing contains ferrite and fine lamellar pearlite. After oil quenching from 880°C and tempering to 630°C the microstructure of the steel is formed by highly tempered martensite. During tempering the hardness decreases from 520 HV to 280 HV and abrasion resistance reduces from 1.78 to 1.38.

The microstructure of soft annealed steels grade 37MnSi5 and 58CrV4 is formed by a mixture of ferrite and spheroidal pearlite. These steels after normalizing have microstructures containing ferrite and lamellar pearlite. Quenching and tempering of these steels produces a microstructure of tempered martensite. The net result of tempering of these steels is a reduction of hardness and abrasion resistance (see Table 4).

Ball bearing steel grade100CrMn6 after spheroidizing has a microstructure of spheroidized pearlite. Oil quenching produces a microstructure containing martensite and carbides. Tempering to 160°C decrease the hardness and abrasion resistance ψ_a (from 2.59 to 2.37). This steel can be used in different applications with intensive abrasive action of mineral particles.

Chromium stainless steel grade X10Cr13 after oil quenching has a martensitic-ferritic microstructure with the hardness 420 HV. Tempering of this steel (up to 450°C) changes the hardness and abrasion resistance in a narrow range.

Chromium stainless steel grade X40Cr13 has a martensitic microstructure after oil quenching owing to a higher content of carbon. It has a higher hardness and abrasion resistance than steel grade X10Cr13. This steel is usually tempered to a low temperature 150°C. During tempering the hardness of the steel decreases, but its abrasion resistance slightly increases.

The microstructure of stainless Cr-Ni or Cr-Ni-Mo steels after austenitizing and water cooling is austenitic with a hardness 166-195 HV. Their abrasion resistance is the same as in quenched and tempered low-alloyed steels (ψ_a = 1.55-1.65).

Austenitic manganese steel grade X120Mn12 (well known Hadfield steel) after austenitizing and water cooling has an austenitic microstructure with an original hardness of 220 HV. With continued impact and /or compression it will surface harden to over 500 HV. It should be noted that only the outer skin surface hardens. The underlayer remains highly ductile and tough. As the surface wears, it continually renews itself becoming harder. During sliding abrasion the work-hardening and the transformation of metastable austenite into strain-induced martensite does not begin. Annealed carbon or low-alloyed steels have the same or higher abrasion resistance (see Table 3 and 4).

5. Abrasion resistance of thermochemical treated structural steels

Using surface treatment on steels can increase abrasion resistance. Chemical composition and microstructure of a surface layer can be altered by surface treatment such as carburizing, nitriding and boronizing, while the bulk microstructure is modified only by transformation hardening. Case-hardening and nitriding were applied to standard mild carbon and low-alloy steels. The chemical composition of the tested steels is presented in Table 2.

The case-hardening of carbon and low-alloy steels comprises the case carburizing to the depth about 1 mm with quenching and tempering at low temperature (150 – 180°C). Usually the surface layer is saturated with carbon to a eutectoid or moderately hypereutectoid content. After quenching the microstructure of surface layer comprises high-carbon martensite and eventually a small amount of cementite. The relative abrasion resistance should therefore be comparable with that of structural or tool steels with a high carbon content.

After carburizing (the thickness of the carburized layer is 0.65 mm), quenching and tempering to a low temperature the microstructure of the carburized layer of chromium-manganese steel grade 20MnCrTi and Cr-Ni grade 17NiCr6 4 steel comprises high-carbon martensite and dispersion cementite particles.

The hardness of case-hardened steels changed in the range of HV 741-754. Relative abrasion resistance ψ_a of tested steels changed from 2.38 to 2.48. The achieved values of ψ_a corresponds to relative abrasion resistance of hardened high carbon steels. During abrasion by an abrasive cloth linear wear was very small. In the case of greater linear wear abrasion resistance drops because the hardness falls due to a carbon content reduction in the subsurface layers. The hardness and physical-mechanical properties deteriorate. The chemical composition of case-hardened steels has a very small influence on their abrasion resistance.

The microstructure of gas nitrided steels is a combination of that of the nitride surface layer and the subsurface diffusion layer. By nitriding at 500°C in gas a nitride layer was formed on both nitriding steels (Cr-Mo-V grade 30CrMoV9 and Cr-Mn-Al grade 33CrAl6) with a thickness of 0.3 mm. The thin surface nitride layer was removed by grinding.

Steel grade	Heat treatment	Hardness (HV)	Abrasion resistance ψ_a
20MnCrTi	C 940°C + Q 820°C/oil + T 150°C/1h/air	754	2.38
17NiCr6 4	C 940°C + Q 820°C/oil + T 150°C/1h/air	741	2.48
33CrAl6	Q 880°C/oil + T 630°C/1h/oil + N 500°C	953	2.05
30CrMoV9	Q 880°C/oil + T 630°C/1h/water + N 500°C	894	1.99

Table 5. Thermochemical treatment of tested structural steels (Q - quenching; T - tempering; C - carburizing; N - nitriding)

In nitriding steels, after removal of the surface nitride layer, surface hardness was in the range 894 - 953 HV which is about 153 - 212 HV higher than in carburized surfaces. The relative abrasion resistance Ψ_a in nitriding steels was 1.99 – 2.05, which is lower than that in carburized low-alloy steels (see Table 5). Nitrided layers have increased abrasion resistance, but their higher hardness is not expressed in higher abrasion resistance in comparison with the surface layers of case-hardened steels. A relative thin surface layer with high hardness does no facilitate their application in the conditions of intensive abrasion. After removing the nitrided layer erosion resistance decreases very quickly.

Table 6 gives for comparison the values of the abrasion resistance of carburized and nitrided steels and carbon steels tested on an apparatus with an abrasive cloth under identical testing conditions. In quenched carbon steels abrasion resistance and also hardness drop markedly at relatively low tempering temperatures. The slightly higher values of hardness and abrasion resistance of quenched grade C60 and C80W2 steels are therefore in good agreement with the values of hardness and abrasion resistance of carburized steels.

Steel grade	Content of carbon (%)	Heat treatment or thermo-chemical treatment	Hardness (HV)	Abrasion resistance ψ_a
20CrMnTi	0.20	carburizing	754	2.38
17NiCr6	0.16	carburizing	741	2.48
33CrAl6	0.36	nitriding	953	2.05
30CrMoV9	0.27	nitriding	894	1.99
C60	0.68	hardened	851	2.46
C80W2	0.85	hardened	874	2.54
C100W2	0.99	hardened	774	1.87

Table 6. Comparison of hardnesses and relative abrasion resistances of carburized, nitrided and selected carbon steels.

In nitriding steels the surface hardness is higher than that in carburized surfaces, however the abrasion resistance is lower. The hardening of surface layers by dispersion nitrides and intersticial nitrogen atoms only insufficiently prevents penetration of the edges and peaks of hard abrasive particles and scratching which occur during the relative motion of the abrasive medium and the nitrided surface.

Boronizing is a thermochemical treatment that diffuses boron through the surface of metallic substrates. On the surface of boronized ferrous alloys, a compound layer is generally developed. This compound layer (or boride layer) is normally composed of two sublayers; the outermost and the innermost are rich in FeB and Fe_2B, respectively (Martini et al., 2004). Boronized steels exhibit high wear resistance owing to the high hardness of the boride layer (typically 1500-1800 HV). For abrasion conditions boronizing is superior to carburizing and nitriding (Wang & Hutchings, 1988; Lin et al., 1990; Atik et al., 2003). But the boride layers are thinner than the carburized or nitrided layers (typically 20 – 120 μm).

6. Abrasion resistance carbon and low-alloyed tool steels

Tool steels are basically medium- to high-carbon steels with specific elements included in different amounts to provide special characteristics. Carbon in the tool steel is provided to help to harden the steel to greater hardness for cutting and wear resistance. Other elements are added to provide greater toughness or strength. These steels were developed to resist wear at temperatures of forming and cutting applications. According to their application they can be dividend into six categories - cold work, shock resisting, hot work, high speed, mold and special purpose tool steels. Mainly cold work tool steels and high speed steels use in conditions of abrasive wear. Chemical composition of several carbon, low-alloyed, ledeburitic and high speed steels is given in the Table 7. These steels were hardened by processes currently recommended by producers (see Table 8). Cold work tool steels are water quenched (carbon steels) or oil quanched (alloyed steels) and tempered in the range of 150 to 250°C. After quenching their microstructure constitutes martensite with some amount of retained austenite. Tempering to low temperatures decreases the hardness and partly eliminates the high internal stresses of these steels.

Steel grade	Content of elements (%)							
	C	Mn	Si	Cr	Ni	Mo	V	W
C45W	0.48	0.80	0.31	-	-	-	-	-
C80W2	0.85	0.24	0.46	-	-	-	-	-
C110W1	1.16	0.29	0.26	-	-	-	-	-
90MnCrV8	0.86	1.90	0.31	0.28	-	-	0.16	-
X185CrWV12	1.85	0.23	0.22	11.93	--	--	0.31	1.47
X195CrVWMo5 4	1.95	0.42	0.38	5.12	--	0.84	4.19	1.50
HS 12-0-4	1.28	0.32	0.09	4.38	--	0.08	3.91	11.85
HS 12-0-2	0.86	0.26	0.25	4.26	--	0.005	2.15	12.00

Table 7. Chemical composition of carbon, low-alloyed, ledeburitic and high speed tool steels

Steel grade	Heat treatment	Hardness (HV)	Abrasion resistance ψ_a
C45W	Q 810⁰C/30min/water	789	1.96
	Q 820⁰C/30min/water + T200⁰C/2h/air	645	1.54
C80W2	Q 790⁰C/30min/water	874	2.54
	Q 800⁰C/30min/water + T200⁰C/2h/air	740	1.85
C110W1	Q 780⁰C/30min/water	889	2.58
	Q 780⁰C/30min/water + T200⁰C/2h/air	740	1.82
90MnCrV8	Q 780⁰C/oil	789	2.27
	Q 780⁰C/oil + T200⁰C/2h/air	670	1.77
	Q 780⁰C/oil + T300⁰C/2h/air	641	1.54

Table 8. Heat treatment of carbon and low-alloy tool steels (A - annealing; Q - quenching; T - tempering)

7. Abrasion resistance of chromium ledeburitic steels

The microstructure of ledeburitic chromium steel X185CrWV12 after soft annealing is ferritic with a large amount of fine and coarse $M_{23}C_6$ and M_3C_7 complex carbides. After quenching (980°C/1h/oil) the microstructure is formed by fine martensite with a large amount of fine spheroidal and coarse carbides. The share of retained austenite is low - when quenching from high temperatures it gradually grows and the share of martensite and carbides decreases. After quenching from 1150°C the microstructure is austenite-carbide with coarser carbides precipitated at grain boundaries. A further increase of the quenching temperature is accompanied by dissolution of carbides and coarsening of austenitic grains. Due to fluctuations of the chemical composition, melting down of grain boundaries and reverse precipitation of ledeburite during cooling can locally occur.

When tempering to 250°C the decrease of hardness is not accompanied by marked structural modifications. When tempering to 400°C the amount of fine carbides slightly increases. Precipitation of carbides and decomposition of retained austenite occur at higher tempering temperatures.

Ledeburitic chromium steel X185CrWV12 is frequently used in conditions of intensive abrasive wear. Abrasion resistance Ψ_a grows with increasing quenching temperatures up to 1100°C, when it reaches its maximum value (Ψ_a=4.13). The hardness of quenched specimens decreases from 801 HV to 621 HV. A further increase of the quenching temperature is accompanied by a decrease of the values of Ψ_a. Tempering in the zone of primary hardness exhibits a decrease of both hardness and abrasion resistance, whereas the decrease is only slight, up to a tempering temperature of 250°C (see Table 9).

The microstructure of ledeburitic Cr-V grade X195CrWVMo5 4 steel after soft annealing comprises ferrite, $M_{23}C_6$ and M_3C_7 complex carbides and special MC carbides. The specimens were oil quenched from a temperature within the range 950 - 1200°C. With a growing temperature gradually the share of martensite decreases, complex and special carbides dissolve so that the contents of carbon and carbide-forming alloy elements in austenite grow. The stability of austenite grows and the positions of the M_s and M_f points shift to lower values. Therefore when quenching from high temperatures the share of retained austenite grows whilst the shares of martensite and carbides decrease.

Ledeburitic Cr-V grade X195CrVWMo5 4 steel belongs to materials with high abrasion resistance. The abrasion resistance Ψ_a grows with a growing quenching temperature and reaches a maximum value for quenching from 1150°C (Ψ_a=6.02). Hardness with a growing temperature drops from 772 HV for quenching from 950°C to 387 HV for quenching at 1200°C. The hardness of the steel quenched from 950°C during tempering decreases, however the decrease is up to 250°C/2h/air only slight. Tempering in the zone of primary hardness also showed a decrease of abrasion resistance. The decrease of hardness and abrasion resistance was marked when tempering to temperatures above 250°C.

In the tested ledeburitic steels alloyed with Cr and Cr-V a positive effect was found of growing quenching temperature on enhancement of abrasion resistance. The maximum value of the relative abrasion resistance was achieved by oil quenching from 1100°C in steel grade X185CrWV12 and from 1150°C in steel grade X195CrVWMo5 4. The relative growth of Ψ_a in steel grade X185CrWV12 after quenching from 1100°C is 26% compared with the value of Ψ_a after quenching from 980°C. The relative growth of abrasion resistance in steel grade X195CrVWMo5 4 after quenching from 1150°C is 42.7% compared with the value of Ψ_a for quenching from 950°C.

Steel grade	Heat treatment	Hardness (HV)	Abrasion resistance ψ_a
X185CrWV12	A 770°C/4h/furnace	216	1.49
	Q 980°C/oil	801	3.28
	Q 980°C/oil + T 150°C/2h/air	774	3.17
	Q 980°C/oil + T 250°C/2h/air	702	2.99
	Q 980°C/oil + T 400°C/2h/air	656	2.66
	Q 980°C/oil + T 500°C/2h/air	639	2.63
	Q 980°C/oil + T 600°C/2h/air	512	2.09
	Q 980°C/oil + T 700°C/2h/air	345	1.61
	Q 1050°C/oil	765	3.48
	Q 1100°C/oil	621	4.13
	Q 1150°C/oil	385	3.84
	Q 1200°C/oil	374	3.58
X195CrVWMo5 4	A 770°C/4h/furnace	265	1.66
	Q 950°C/oil	772	4.22
	Q 950°C/oil + T 150°C/2h/air	754	3.93
	Q 950°C/oil + T 250°C/2h/air	770	3.84
	Q 950°C/oil + T 400°C/2h/air	685	2.98
	Q 950°C/oil + T 500°C/2h/air	650	2.85
	Q 950°C/oil + T 600°C/2h/air	512	2.40
	Q 950°C/oil + T 700°C/2h/air	332	1.64
	Q 1000°C/oil	730	4.50
	Q 1050°C/oil	723	5.00
	Q 1100°C/oil	523	5.73
	Q 1150°C/oil	454	6.02
	Q 1200°C/oil	387	5.59

Table 9. Heat treatment of ledeburitic chromium steels (A - annealing; Q - quenching; T - tempering)

The decisive role in increasing the abrasion resistance with growing quenching temperatures is assigned to the growing volume share of retained austenite, which, during plastic deformation caused by penetration of abrasive particles and scratching of the surface subject to wear, is hardened and partially transforms to deformation martensite (Berns et al., 1984; Suchánek & Bakula, 1987). The positive effect of the growing volume share of retained austenite in the microstructure of ledeburitic chromium and chromium-vanadium steels is related with the growth of energy required for the separation of the unit volume of particles. The sum of energies for phase transformation, internal stresses and for creation of a critical dislocation density is much higher than the decrease of energy required for penetration of abrasive particles into the surface subject to wear (decrease of hardness) and energy required for the break-up of carbides (decrease of volume share of carbides in the microstructure). Modifications of the microstructure of ledeburitic steels caused by the growth of quenching temperatures are in Table 10.

Steel grade	Heat treatment	Volume share of retained austenite (%)	Surface share of carbides (%)	Type of carbides
X185CrWV12	A 770°C/4h/furnace	0	17.1	M_7C_3
	Q 980°C/1h/oil	18.7	13.7	M_7C_3
	Q 1050°C/1h/oil	36.1	12.65	M_7C_3
	Q 1100°C/1h/oil	45.9	11.56	M_7C_3
	Q 1150°C/1h/oil	89.4	5	M_7C_3
	Q 1200°C/1h/oil	97	2.1	M_7C_3
X195CrWVMo 5 4	A 770°C/4h/furnace	0	13.95	MC (M_7C_3, $M_{23}C_6$)
	Q 950°C/1h/oil	15.2	8.29	MC (M_7C_3)
	Q 1000°C/1h/oil	24.2	6.04	MC (M_7C_3)
	Q 1050°C/1h/oil	34.5	3.99	MC
	Q 1100°C/1h/oil	53.3	3.4	MC
	Q 1150°C/1h/oil	72.7	2.37	MC
	Q 1200°C/1h/oil	92.5	1.37	MC

Table 10. Volume share of retained austenite, surface share and type of carbides in ledeburitic steels

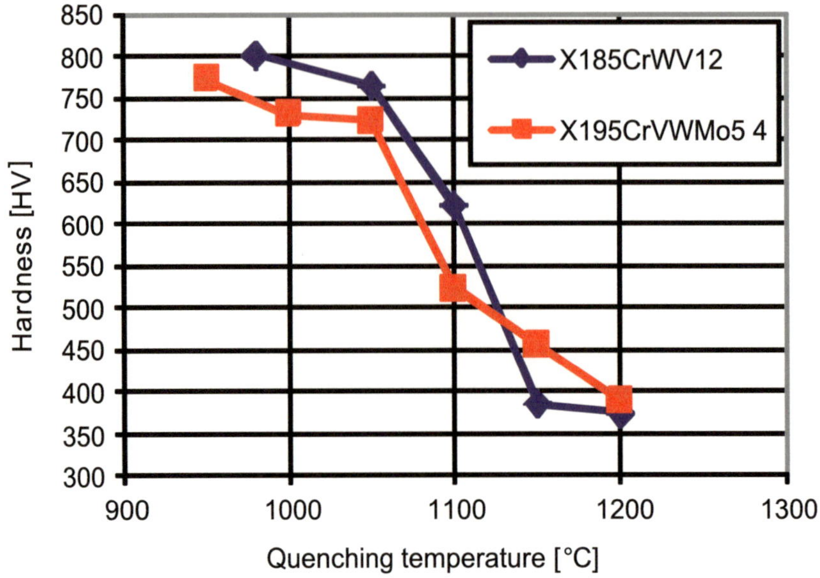

Fig. 3. Effect of quenching temperature on hardness of ledeburitic steels

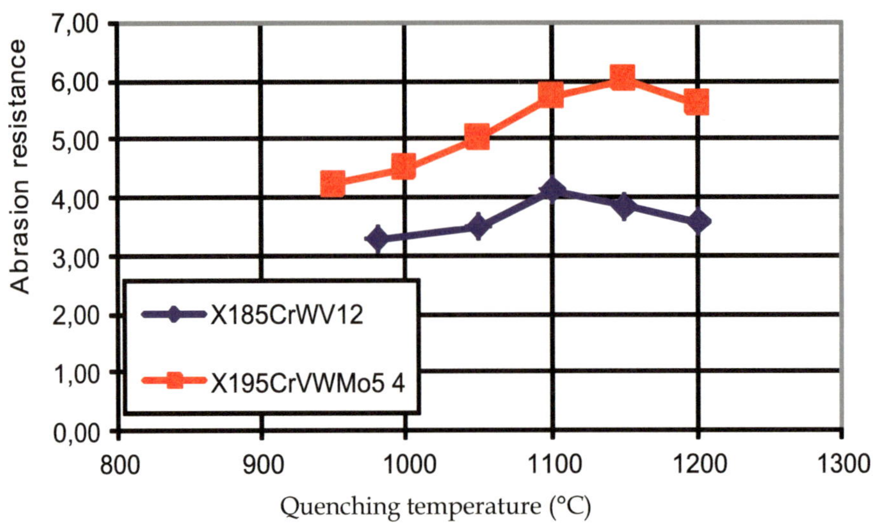

Fig. 4. Abrasion resistance of of ledeburitic steels vs. quenching temperature

The decrease of abrasion resistance with growing tempering temperatures is related with structural and phase modifications which occur both in martensite nad retained austenite. Due to the fact that the value of Ψ_a dropped after quenching to 250°C/2h/air only by 9%, it can be assumed that internal stress relief and creation of zones enriched with carbon which gradually take the character of coherent or semicoherent particles do not considerably decrease abrasion resistance.

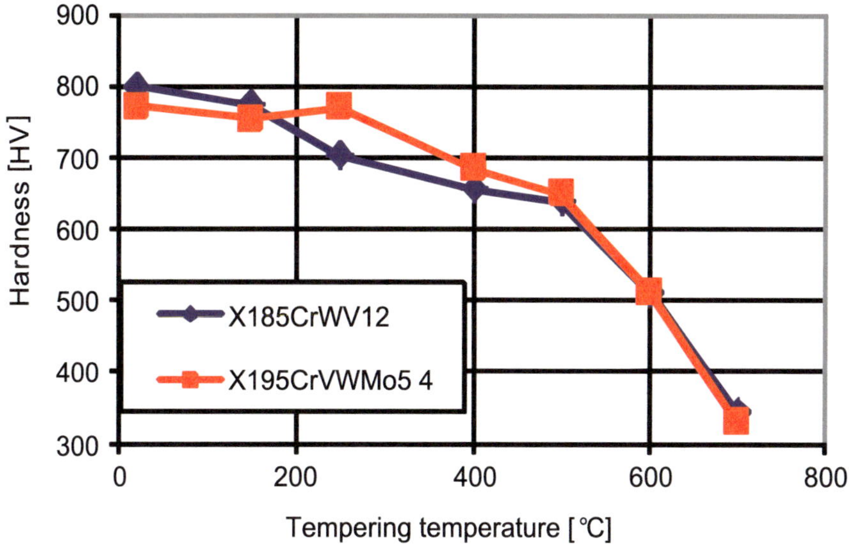

Fig. 5. Hardness of ledeburitic steels vs.tempering temperature

A marked drop of Ψ_a appeared at a tempering temperature of 400°C. This is related with a drop of the concentration of carbon in martensite and in retained austenite and creation of ε-carbides. A decrease of the carbon content in martensite impairs its physical and mechanical properties in spite of the fact that formation of ε-carbides causes hardening. Another factor which decreases abrasion resistance is the anisotropic decomposition of retained austenite. Tempering to temperatures above 400°C leads to a marked decrease of resistance to wear and also hardness (see Fig. 5 and 6).

Values of abrasion resistance after tempering to temperatures in the zone of primary hardness drop with growing tempering temperatures. Ledeburitic Cr and Cr-V steels should preferably be tempered to a maximum temperature of 250°C/2h/air.

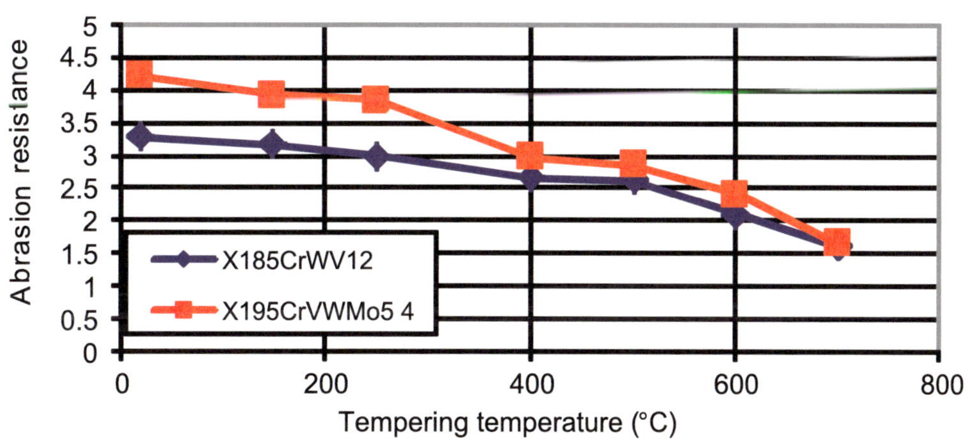

Fig. 6. Abrasion resistance Ψ_a of ledeburitic steels vs. tempering temperature

8. Abrasion resistance of high speed steels

The microstructure of high-speed grade HS 11-0-4 steel after soft annealing is pearlite-carbide. The microstructure after quenching comprises martensite, retained austenite and carbides. With growing temperatures within the range 1000 - 1270°C M_6C and $M_{23}C_6$ complex carbides and MC special carbides (at temperatures >1200°C) dissolve. With an increasing content of carbon and carbide forming elements the hardness of martensite grows and at high quenching temperatures also the content of retained austenite (up to 20 - 25%). By tempering in the zone of primary hardness internal stresses are relieved, carbon diffuses from the martensite lattice, ε-carbides are formed and at higher tempering temperatures carbon precipitation and decomposition of retained austenite occurs. In the zone of secondary hardness at repeated tempering, anisotropic decomposition of retained austenite, tempering of martensite and strengthening of the martensitic matrix by precipitated type M_6C, $M_{23}C_6$ and MC carbides occurs.

The microstructure of high-speed grade HS 11-0-2 steel oil quenched from temperatures in the range 1200 - 1270°C comprises martensite, retained austenite, and M_6C, $M_{23}C_6$ complex carbides and special MC carbides. With growing quenching temperatures the share of carbides and martensite decreases and the share of retained austenite grows. In the zone of secondary hardness at repeated tempering the following processes occur: anisotropic decomposition of retained austenite, tempering of martensite and hardening of the martensitic matrix by precipitated secondary carbides. Hardnesses achieved by tempering in the zone of maximum secondary hardness are in many cases higher than those achieved after quenching.

Abrasion resistance in high-speed grade HS11-0-4 steel grows with a growing quenching temperature and slightly decreases in the zone of primary hardness with tempering temperatures growing up to 250°C (see Table 10). The same behaviour of abrasion resistance was found in high-speed grade HS11-0-2 steel. A high increase of hardness and abrasion resistance with growing quenching temperatures is related with structural modifications in high-speed steels. Carbides dissolve and the carbon content and share of carbide-forming elements in austenite grows. This is manifested by greater stability during quenching. The share of retained austenite grows. The highest values of Ψ_a (Ψ_a=6.88) were found in grade HS 11-0-4 steel for quenching from 1270°C, i.e. from the highest point of the range of quenching temperatures recommended by the producer of the steel (see Table 11).

The decrease of abrasion resistance in the zone of primary hardness is more marked in steels quenched from higher temperatures. When tempering in the zone of secondary hardness, both the hardness and the values of Ψ_a grow with rising quenching temperatures. Maximum abrasion resistance and hardness of quenched and tempered specimens were achieved after three-fold repeated tempering to 560°C. Tempering to temperatures above the maximum of secondary hardness leads to a marked decrease of both hardness and abrasion resistance (see Table 9).

High-speed steels can be considered as composite materials where large primary carbides (1 - 10 μm) are dispersed in the martensitic matrix together with a much finer dispersion of small secondary carbides (<100 nm). Secondary carbides are the cause of precipitation hardening of the martensitic matrix. Primary carbides, namely MC and M_6C (1500 - 2800 HV) are harder than the matrix and consequently enhance abrasion resistence (Suchánek & Průcha, 1984; Badisch & Mitterer, 2003; Wei et al., 2006).

Steel grade	Heat treatment	Hardness (HV)	Abrasion resistance ψ_a
HS11-0-4	A 800°C/4h/furnace	223	1.25
	Q 1000°C/oil	578	1.98
	Q 1050°C/oil	636	2.34
	Q 1050°C/oil + T 150°C/1h/air	753	2.39
	Q 1100°C/oil	743	3.22
	Q 1100°C/oil + T 150°C/1h/air	800	2.84
	Q 1100°C/oil + T 250°C/1h/air	698	2.86
	Q 1150°C/oil	826	3.90
	Q 1150°C/oil + T 150°C/1h/air	801	3.70
	Q 1150°C/oil + T 250°C/1h/air	685	3.52
	Q 1150°C/oil + T 400°C/1h/air	654	3.44
	Q 1170°C/oil + 3×T 610°C/45′/air	740	4.27
	Q 1200°C/oil + T 150°C/1h/air	772	4.31
	Q 1200°C/oil + T 250°C/1h/air	675	3.60
	Q 1210°C/oil	853	6.03
	Q 1210°C/oil + 3×T 560°C/45′/air	843	4.60
	Q 1210°C/oil + 3×T 600°C/45′/air	739	3.72
	Q 1210°C/oil + 3×T 640°C/45′/air	653	3.04
	Q 1240°C/oil	848	6.10
	Q 1240°C/oil + 3×T 150°C/45′/air	828	4.56
	Q 1240°C/oil + 3×T 250°C/45′/air	732	3.74
	Q 1240°C/oil + 3×T 400°C/45′/air	728	3.68
	Q 1240°C/oil + 3×T 500°C/45′/air	757	4.56
	Q 1240°C/oil + 3×T 560°C/45′/air	882	5.15
	Q 1240°C/oil + 3×T 600°C/45′/air	761	4.04
	Q 1240°C/oil + 3×T 640°C/45′/air	692	2.94
	Q 1240°C/oil + 3×T 680°C/45′/air	598	2.25
	Q 1270°C/oil	802	6.88
	Q 1270°C/oil + 3×T 560°C/45′/air	900	6.33
HS11-0-2	Q 1200°C/oil	709	2.59
	Q 1200°C/oil + 3×T 560°C/45′/air	791	3.06
	Q 1240°C/oil	758	2.89
	Q 1240°C/oil + 3×T 560°C/45′/air	852	3.34
	Q 1240°C/oil + 3×T 570°C/45′/air	887	2.92
	Q 1240°C/oil + 3×T 600°C/45′/air	766	2.91
	Q 1240°C/oil + 3×T 640°C/45′/air	671	2.57
	Q 1240°C/oil + 3×T 670°C/45′/air	587	2.14
	Q 1270°C/oil	752	3.31
	Q 1270°C/oil + 3×T 560°C/45′/air	879	3.63

Table 11. Heat treatment, hardness and abrasion resistance of high-speed steels (A – annealing; Q – quenching; T – tempering)

Enhancement of hardness and abrasion resistance with growing quenching temperatures is related with structural modifications of high-speed steels. M_6C and $M_{23}C_6$ complex carbides dissolve at quenching temperatures and special MC carbides dissolve at temperatures above 1200°C. The presence of higher carbon and carbide forming additions induces growth of the strength and hardness of martensite and at high temperatures also growth of the volume share of retained austenite (up to 20-25%). The higher strength of the martensitic matrix manifests itself positively by a higher resistance to scratching by abrasive Al_2O_3 particles. Besides the matrix also special MC carbides with hardnesses in the range 2700-2990 HV play a significant role - they blunt the cutting edges of abrasive particles and decrease their scratching effect. The decrease of macrohardness after quenching from 1270°C was not accompanied by a decrease of relative abrasion resistance. It can be assumed that hardening of retained austenite and its partial transformation to martensite initiated by the penetration of edges of abrasive particles and their scratching effect has eliminated the drop of the share of carbides and grain coarsening. During tempering in the zone of primary hardness (150°C/30′/air) the following processes occur : internal stress relief, diffusion of carbon from the martensite lattice and formation of ε-carbides. These processes proceed with a higher intensity after quenching from higher temperatures and hence the decrease of Ψ_a values is more pronounced.

A further drop of the values of abrasion resistance during tempering at 250°C and/or 400°C is less pronounced since the matrix is hardened by fine particles.

A marked maximum of hardness and abrasion resistance during tempering in the zone of maximum secondary hardness (560°C/3x45′) is closely related with the anisotropic decomposition of retained austenite and hardening of the matrix by precipitated M_6C, $M_{23}C_6$ and MC carbides [25]. High-speed grade HS 11-0-2 steel has a lower content of carbon and vanadium and hence also a lower share of carbides. Compared with high-speed grade HS 11-0-4 steel it has a markedly lower abrasion resistance. The high content of vanadium both in high-speed and ledeburitic steels positively affects abrasion resistance.

9. Conclusions

1. Abrasion resistance of carburized low-alloy steels is on the same level as in high-carbon structural and tool steels.
2. Resistance of nitrided steels to abrasion by hard particles is lower than that of carburized low-alloy steels.
3. Relative abrasion resistance of ledeburitic chromium and high-speed steels grows with growing quenching temperatures. In ledeburitic chromium grade X185CrWV12 steel maximum abrasion resistance was achieved by quenching from 1100°C whilst in ledeburitic chromium-vanadium steel the optimum quenching temperature was 1150°C. Maximum abrasion resistance in high speed steels was achieved by quenching from 1270°C.
4. Values of relative abrasion resistance after tempering to temperatures in the zone of primary hardness drop with growing tempering temperatures. Ledeburitic Cr and Cr-V steels should preferably be tempered to a maximum temperature of 250°C/2h/air.
5. The values of abrasion resistance in high-speed steels in the zone of secondary hardness grow with growing quenching temperatures. The values of Ψ_a achieved are lower than those obtained only by quenching.

6. Relative abrasion resistance in high-speed steels quenched and tempered in the zone of secondary hardness drops with growing tempering temperatures. The optimum tempering temperature is 560°C/3x45´/air.

10. References

Atik, E., Yunker, U. & Meric, C. (2003). The effects of conventional heat treatment and boronizing on abrasive wear and corrosion of SAE 1010, SAE 1040, D2 and 304 steels. *Tribology Intern.*, Vol. 36, pp. 155-161, ISSN 0301-679X

Badisch, E., & Mitterer, C. (2003). Abrasive wear of high speed steels: Influence of abrasive particles and primary carbides on wear resistance. *Tribology Intern.*, Vol. 36, pp. 765-770, ISSN 0301-679X

Berns, H., Gümpel, P., Trojahn, W. & Weigand, H. H. (1984). Gefüge und abrasiver Verschleiß von Kaltarbeitsstählen mit 12% Cr. *Archiw f. Eisenhüttenwesen*, Vol. 55, No. 6, pp. 267-270

Eyre, T. S. (1979). *Wear Resistance of Metals.* In: *Treatise on Materials Science and Technology*, vol. 13, Wear, Academic Press, Inc., pp. 363-442

Filippova, L. T. & Goldshtein, J. J. (1979). Effect of composition and structure on the resistance of steels to abrasive wear. *Metalloved. Term. Obrab. Met.*, No. 2, pp. 10- 15, ISSN 0026-0673 (in Russian)

Fremunt, P., Musilova, L. & Pacal, B. (1971) Effect of V on the increase in abrasion resistance of ledeburitic Cr steels. *Slévárenství*, Vol. 19, No. 11, pp. 466-471 (in Czech)

Grinberg, N. A. Livshits, L. S. & Sherbakova, V. S.(1971). Influence of alloying ferrite and carbide phase on wear resistance of steels. *Met Sci Heat Treat.*, Vol. 13, No. 9-10, pp. 768-770 ISSN 0026-0673

Krushchov, M. M. & Babichev, M. A. (1960). *Research of metal wear*, Publ. House AN SSSR, Moscow, (in Russian)

Krushchov, M. M. & Babichev, M. A. (1970). *Abrasive wear*. Publ. House Science (Nauka), Moscow, (in Russian)

Khruschov, M. M. (1974). Principles of abrasive wear. *Wear*, Vol. 28, pp. 69-88, ISSN 0043-1648

Larsen-Badse, J. & Mathew, K. G. (1966). Abrasion of Some Hardened and Tempered Carbon Steels. *Trans. Metall. Soc. AIME*, Vol. 236, pp. 1461-1466

Lin, Z., Wang, Z. & Sun, X. (1990). The influence of internal stress and preffered orientation on the abrasive wear resistance of a boronized medium carbon steel. *Wear*, Vol. 138, pp. 285-294, ISSN 0043-1648

Martini, C., Palombarini, G., Poli, G. & Prandstraller, D. (2004). Sliding and abrasive wear behaviour of boride coatings. *Wear*, Vol. 256, pp. 608–613 ISSN 0043-1648

Moore, M.A. (1974). The relationship between the abrasive wear resistance, hardness and microstructure of ferritic materials. *Wear*, Vol. 28, pp. 59-68, ISSN 0043-1648

Petrov, I.V. & Grinberg, N.A. (1968). Wear resistance of hardfacing under abrasive wear conditions. *Svar. Proizv.*, No. 2, pp. 30-33 (in Russian)

Popov, V. S., Nagornyj, P.L. & Garbuzov, A. S. (1969). Specific energy of rupture of carbides and the wear resistance of alloys. *Fizika Metallov Metallovedenie*, Vol. 28, No.2, pp. 332–336 (in Russian)

Popov, V. S. & Nagorny, P.L. (1969). Influence of carbides on the abrasive wear resistance of alloys. *Liteinoe Proizv.*, No. 8, pp. 27-29 (in Russian)

Popov, V.S. & Nagorny, P.L. (1970). Abrasive wear of chromium alloys. *Liteinoe Proizv.*, No. 3, pp. 27-28 (in Russian)

Richardson, R. C. D. (1967). The wear of metal by hard abrasives. *Wear,* Vol. 10, pp. 353-382, ISSN 0043-1648

Roberts, G. A., Hamaker, J. C. & Johnson, A. R. (1962). *Tool Steels*. ASM, Metal Park, Ohio

Schabuyev, S. A., Mrkrtychan, Y. S & Pishchanskiy, V. I.(1972). Influence of composition and structure of chromium alloys on their abrasion resistence. *Liteinoe Proizv.,* No. 3, pp. 28-29 (in Russian)

Simm, W. & Freti, S. (1989) Abrasive wear of multiphase materials. *Wear,* Vol.129, pp.105-121, ISSN 0043-1648

Soroko-Novickaya, A. A. (1959). Wear resistance of carbon steel with different structure. *Friction and Wear in Machines (Trenie i iznos v maschinach)*, Vol. 13, pp. 5-18, Publ. House AN SSSR, Moscow.

Suchánek, J. & Průcha, J.(1984) The effect of thermal treatment and structure on the resistance to abrasive wear of the high speed cutting steel 19 810. In: *Proc. 2-nd Symposium „INTERTRIBO 84"*, pp. 678-682, High Tatras, October 3-5,1984 (in Czech)

Suchánek, J. & Bakula, J. (1987). Influence of structural factors on resistance of metallic materials in abrasive wear. *Strojírenství*, Vol. 37, pp. 232-236 (in Czech)

Suchánek, J., Kuklík, V. & Zdravecká, E. (2007). *Abrazívní opotřebení materiálů (Abrasive wear of materials)*. CTU Prague, ISBN 978-80-01-03659-4 (in Czech)

Tsypin, I.I. (1983). *Wear resistant white cast irons. Structure and properties*. Moscow, Metallurgiya (in Russian)

Uetz, H. (1986). *Abrasion and Erosion,* Carl Hanser Verlag, München, Wien

Vocel, M. & Dufek, V. (1976). *Friction and wear of machine parts*. SNTL, Prague, 374 pp. (in Czech)

Wang, A. G. & Hutchings, I. M. (1988). Mechanisms of abrasive wear in a boronized alloy steel. *Wear*, Vol. 124, pp. 149-163, ISSN 0043-1648

Wei, S. , Zhu, J. & Xu, L. (2006). Effect of vanadium and carbon on microstructures and abrasive wear resistance of high speed steel. *Tribology Intern.*, Vol. 39, pp. 641-648, ISSN 0301-679X

Wirojanupatump, S. & Shipway, P. H. (1999). A direct comparison of wet and dry abrasion behaviour of mild steel. *Wear,* Vol. 223-235, pp. 655-665, ISSN 0043-1648

Zum Gahr, K.H. (1985). Abrasive wear of two–phase metallic materials with a coarse microstructure. In: K.C. Ludema, (Ed.), *International Conference on Wear of Materials*, ASME, Vancouver, p. 793

7

Analysis of Abrasion Characteristics in Textiles

Nilgün Özdil[1], Gonca Özçelik Kayseri[2] and Gamze Süpüren Mengüç[2]
[1]Ege University, Textile Engineering Department, Izmir,
[2]Ege University, Emel Akın Vocational Training School, Izmir,
Turkey

1. Introduction

Wear in textile materials is one of a limited number of fault factors in which an object loses its usefulness and the economic implication can be of enormous value to the industry. The terms wear and abrasion are sometimes confused. Wear is a very general term covering the loss of material by virtually any means. As wear usually occurs by rubbing together of two surfaces, abrasion is often used as a general term to mean wear (Brown, 2006).

The resistance of textile materials to abrasion as measured on a testing machine in the laboratory is generally only one of several factors contributing to wear performance or durability as experienced in the actual use of the material. While "abrasion resistance" (often stated in terms of the number of cycles on a specified machine, using a specified technique to produce a specified degree or amount of abrasion) and "durability" (defined as the ability to withstand deterioration or wearing out in use, including the effects of abrasion) are frequently related, the relationship varies with different end uses, and different factors may be necessary in any calculation of predicted durability from specific abrasion data (ASTM D 4966).

Abrasion is the physical destruction of fibres, yarns, and fabrics, resulting from the rubbing of a textile surface over another surface (Abdullah et al., 2006). Textile materials can be unserviceable because of several different factors and one of the most important causes is abrasion. Abrasion occurs during wearing, using, cleaning or washing process and this may distort the fabric, cause fibres or yarns to be pulled out or remove fibre ends from the surface (Hu, 2008; Kadolph, 2007). Abrasion ultimately results in the loss of performance characteristics, such as strength, but it also affects the appearance of the fabric (Collier & Epps, 1999).

The main factors that reduce service life of the garment are heavily dependent on its end use. But especially certain parts of apparel, such as collar, cuffs and pockets, are subjected to serious wear in use (Figure 1). Abrasion is a serious problem for home textiles like as carpets and upholstery fabrics, socks and technical textiles as well. Yarn abrasion is another important subject that should be considered during processing.

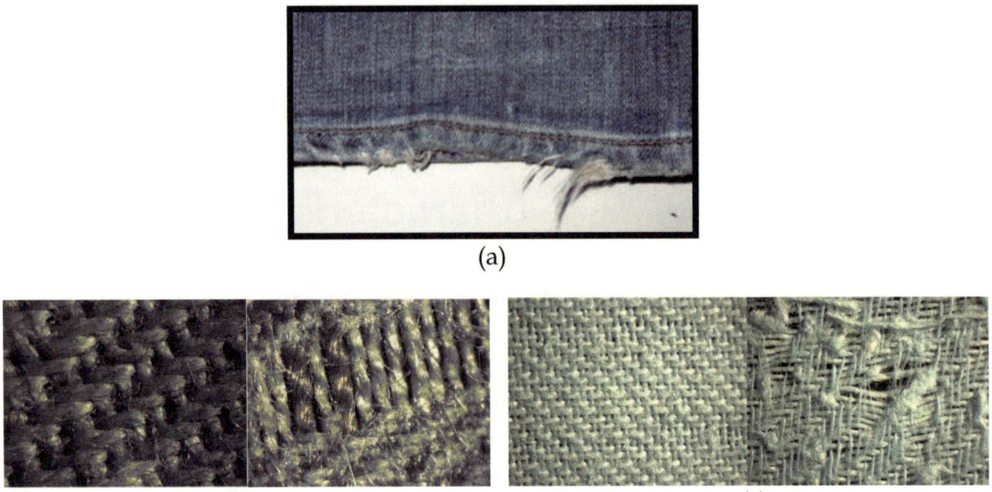

(a) (b) (c)

Fig. 1. Abraded textile products (a) edge of pants, (b) (c) surface appearances of fabrics - before (left) and after (right) the abrasion test.

In this chapter, detailed information about the abrasion and abrasion resistance of the textile materials are discussed. In the first part, the abrasion and wear mechanism are explained. In the second part, abrasion of the fabrics (factors affecting abrasion such as fibre, yarn and fabric properties, parameters affecting the test results, testing and evaluation methods), yarn abrasion (yarn on yarn and yarn external abrasion), abrasion characteristics of socks and technical textile fabrics are analyzed. Studies on the mentioned subjects are given as well.

2. Abrasion mechanism of textiles

Abrasive wear in textiles is caused by different conditions mainly given below:

- Friction between textile materials, such as rubbing of a jacket or coat lining on a shirt, pants pockets against pants fabric etc.
- Friction between the textile materials to the external object, such as rubbing of trousers to the seat, friction of the yarn to the needle etc.
- Friction between the fibres and dust, or grit, in a fabric that results in cutting of the fibres. This is an extremely slow process, it may be observed on flags hanging out or swimwear because of the unremoved sand.
- Friction between the fabric components. Flexing, stretching, and bending of the fibres during the usage causes fibre slippage, friction to each other and breakage (Mehta, 1992).

The study of the processes of wear is part of the discipline of tribology and the mechanism of wear is very complex. Under normal mechanical and practical procedures, the wear-rate normally changes through three different stages: primary stage or early run-in period, where surfaces adapt to each other and the wear-rate might vary between high and low; secondary stage or mid-age process, where a steady rate of wearing is in motion. Most of the components operational life is comprised in this stage. Tertiary stage or old-age period is where the components are subjected to rapid failure due to a high rate of wearing (http://en.wikipedia.org/wiki/Wear). Some commonly referred to wear mechanisms include: Adhesive wear, abrasive wear, surface fatigue, fretting wear, erosive wear. Adhesive, abrasive wear and surface fatigue mechanism play an important role in the abrasion mechanism of the yarns and fabrics.

Adhesive wear, occurs between surfaces during frictional contact and generally refers to unwanted displacement and attachment of wear debris and material compounds from one surface to another. The adhesive wear and material transfer due to direct contact and plastic deformation are the main issues in adhesive wear. The asperities or microscopic high points or surface roughness found on each surface, define the severity on how fragments of oxides are pulled off and adds to the other surface. This is partly due to strong adhesive forces between atoms, but also due to accumulation of energy in the plastic zone between the asperities during relative motion.

Abrasive wear, occurs when a hard rough surface slides across a softer surface. ASTM (American Society for Testing and Materials) defines it as the loss of material due to hard particles or hard protuberances that are forced against and move along a solid surface. Abrasive wear is commonly classified according to the type of contact and the contact environment. The type of contact determines the mode of abrasive wear. The two modes of abrasive wear are known as two-body and three-body abrasive wear. Two-body wear occurs when the grits or hard particles remove material from the opposite surface. The common analogy is that of material being removed or displaced by a cutting or plowing operation. Three-body wear occurs when the particles are not constrained, and are free to roll and slide down a surface.

Fatigue wear of a material is caused by a cycling loading during friction. Fatigue occurs if the applied load is higher than the fatigue strength of the material. Fatigue cracks start at the material surface and spread to the subsurface regions. The cracks may connect to each other resulting in separation and delamination of the material pieces. One of the types of fatigue wear is fretting wear caused by cycling sliding of two surfaces across each other with small amplitude (oscillating). The friction force produces alternating compression-tension stresses, which result in surface fatigue (http://en.wikipedia.org/wiki/Wear, June2011).

In terms of wear mechanism in textiles, abrasion first modifies the fabric surface and then affects the internal structure of the fabric, damaging it (Manich et.al, 2001; Kaloğlu et al., 2003). Good abrasion resistance depends more on a high energy of rupture than on high tenacity at break. Abrasion is not influenced so much by the energy absorbed in the first deforming process (total energy of rupture), as by the work absorbed during repeated deformation. This work is manifested in the elastic energy or the recoverable portion of the total energy. Thus, to prevent abrasion damage, the material must be capable of absorbing energy and releasing that energy upon the removal of load. Energies in tension, shear, compression, and bending are all important for the evaluation of surface abrasion; however, these energies are unknown, and therefore elastic energies in tension permit at least a quantitative interpretation of abrasive damage in fibres and fabrics (Abdullah et al., 2006; as cited in Hamburger, 1945).

Fibres in use are subject to a variety of different forces, which are repeated many times (Hearle & Morton, 2008). The gradual breakdown of the internal cohesion of the individual fibres or by a gradual breakdown of the forces of structural cohesion between the fibres results fabric failure. The relative occurrence of these two phenomena depends to a great extent upon the fabric geometry, but there are limitless factors involved (e.g., construction of yarns and weaves) depending on the individual behavior of different fibres (Figure 2) (Abdullah et al., 2006).

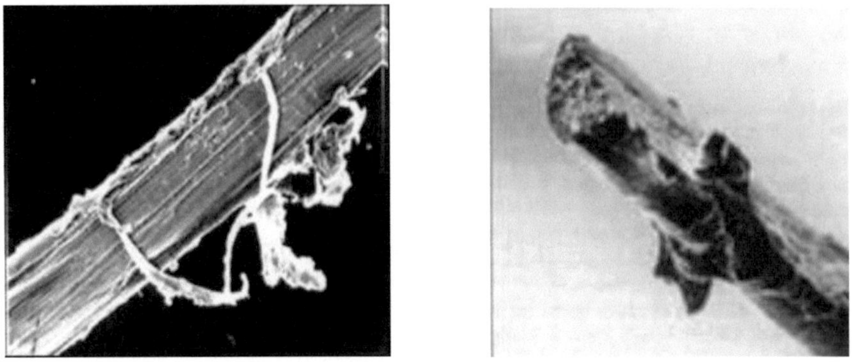

(a) (b)

Fig. 2. Abrasion of fibres over rotating pin: (a) nylon, (b) wool (Hearle & Morton, 2008)

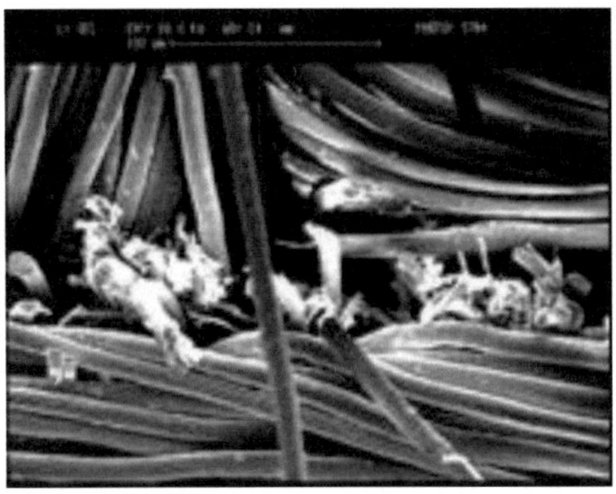

Fig. 3. Fibre rupture occurred as abraded against standard worsted fabric (Abdullah et al., 2006)

During the course of abrasion in textiles, fibre to fibre cohesion plays an important role, usually influenced by yarn twist or close fibre packing. Abrasion behavior indicates that fibre cohesion is strong in the fabric system, and it causes the shear of the fibres themselves. Frictional forces developed in the yarn due to the motion of the abrasion test are dissipated largely in the fibres by the development of tensile and shear stresses; repetition of such stresses results in fibre fatigue, which causes the loss of fibre mechanical properties, leading to rupture. Fibres in the crowns are broken down in succession, and this causes a reduction in fibre cohesion and yarn strength. In lateral abrasion cycles, frictional forces are able to displace fibre from their normal position, and these fibres ruptures through bending and flexing (Figure 3). In addition, it is also possible that some cracking is initiated by abrasion and then propagated by bending action (Abdullah et al., 2006).

Although the mechanism of abrasion is similar, because of the differences in the measurement methods, abrasion resistance of textile materials can be studied in two parts such as fabric abrasion and yarn abrasion

3. Abrasion of fabrics

3.1 Factors affecting abrasion resistance of fabrics

Abrasion resistance of the textile materials is very complex phenomenon and affected by many factors, mainly classified as follows: Fibre, yarn, fabric properties and finishing processes. Some of these parameters affect fabric surface whereas some of them has an influence on internal structure of the fabrics. For example fibre characteristics like wool ratio and fineness play a significant role in surface abrasion, while yarn and fabric characteristics like yarn linear density and interlacing coefficient are significantly related with structural abrasion (Manich et al., 2001).

3.1.1 Fibre properties

The mechanical properties and dimensions of the fibres are important for abrasion. Fibre type, fibre fineness and fibre length are the main parameters that affect abrasion.

Fibres with high elongation, elastic recovery and work of rupture have a good ability to withstand repeated distortion; hence a good degree of abrasion resistance is achieved. Nylon is generally considered to have the best abrasion resistance, followed by polyester, polypropylene (Hu, 2008). Blending either nylon or polyester with wool and cotton is found to increase abrasion resistance at the expense of other properties (Saville, 1999). Higher wool rate increase the mass loss (Manich et al., 2001). Acrylic and modacrylic have a lower resistance than these fibres while wool, cotton and high modulus viscose have a moderate abrasion resistance. Viscose and acetates are found to have the lowest degree of resistance to abrasion. However, synthetic fibres are produced in many different versions so that the abrasion resistance of particular variant may not conform to the general ranking of fibres (Saville, 1999).

The removal of the fibres from yarn structure is one of the reasons of the abrasion. Therefore factors that affect the cohesion of yarns will influence the abrasion resistance of fabrics as well. Longer fibres incorporated into a fabric confer better abrasion resistance than short fibres because it is harder to liberate them from the fabric structure. For the same reason filament yarns are more abrasion resistant than staple yarns made from the same fibre (Saville 1999; Hu, 2008).

The using of finer fibres in the production of yarns causes increment in the number of the fibre in cross section with higher cohesion which results better abrasion resistance. So abrasion retention is better for fabrics with finer fibres (Kaloğlu et al., 2003).

3.1.2 Yarn properties

Yarn structure, count, twist and hairiness are the main properties which affect abrasion of the textile fabrics. Increasing linear density at constant fabric mass per unit area increases the abrasion resistance of the fabrics (Saville, 1999). As yarn got thinner, abrasion resistance values of knitted fabrics decrease and breaking occurs in lower cycles (Özgüney et al., 2008).

Twist is another parameter affecting abrasion. There is an optimum amount of twist in a yarn to give the best abrasion resistance. At low-twist, fibres can easily be removed from the yarn so that it is gradually reduced in diameter. At high twist levels the fibres are held more tightly but the yarn is stiffer so it is unable to distort under pressure when being abraded (Saville, 1999).

Yarn hairiness has a negative effect in terms of mass loss during abrasion. An increase in yarn hairiness, due to the higher level of protruding fibres from yarn surface, reduces fabric abrasion resistance.

The production method of yarn has also an influence on the abrasion resistance, such that carded fabric gives lower resistance than that of combed fabric (Manich et al., 2001). Even yarn structure, using long fibre and lower yarn hairiness are the reasons of that result. Knitted fabrics from ring spun yarns have better abrasion resistance than knitted fabrics from OE spun yarns (Candan et al.,2000; Candan & Önal, 2002). Ring spun yarns are hairier but more compactly structured than OE yarns, this well aligned compact structure doesn't promote easy fibre wear off (Paek, 1989).

Compact yarn fabrics have higher abrasion resistance values compared to the ring yarn fabrics with the same fabric construction. Since the fibres of compact yarns are held more tightly within the yarn structure and higher participation of the fibres into the yarn structure exists, compact yarns have a denser and closer structure compared to the ring yarns. The compact yarn has lower hairiness, high tensile resistance as a result of that fibre movements causing limited abrasion (Akaydın, 2009, 2010). Fabrics woven from compact yarns have also lower weight loss compared to those woven from ring yarns (Ömeroglu & Ülkü, 2007). Sirospun is a modified ring spinning process that two rovings per spindle are fed to the drafting system within specially developed condensers separately and drafted simultaneously. Fabrics knitted from sirospun yarns show better abrasion resistance than ring, air-jet and OE yarns because of the better evenness, hairiness, regular and tightly structure (Örtlek et al., 2010). However as considering the results of fabrics produced with two ply yarns, fabrics from sirospun yarns wear faster than two fold ring spun yarn (Kaloğlu et al., 2003).

Another factor that affects the abrasion is the number of yarn plies. As the number of ply threads per yarn increases, the thickness and the mass per unit area increases and it causes an improvement in abrasion characteristics of the fabric.

3.1.3 Fabric properties

Fabric construction, thickness, weight, the number of yarn (thread density) and interlacing per unit area are the fabric properties affecting abrasion.

Weave type has a significant effect on abrasion resistance of the fabrics. Woven fabric properties will differ by changing the weave pattern which is evaluated not only as an appearance property, but also as a very important structure parameter. If one set of yarns is predominantly on the surface then this set will wear most; this effect can be used to protect the load bearing yarns preferentially. Long yarn floats and a low number of interlacings cause the continuous contact area of one yarn strand to expand and this facilitates the yarn to lose its form more easily by providing easier movement as a result of the rubbing motion. So long floats in a weave such as sateen structures are more exposed and abrade faster, usually cause breaking of the yarns and increasing the mass loss. In this way, holding the fibres in the yarn structure becomes harder and the removal of fibre becomes easier (Kaynak & Topalbekiroğlu, 2008). But the fabrics that have lower floats such as flat plain weave fabrics have better abrasion resistance than other weaves because the yarns are more tightly locked in structure and the wear is spread more evenly over all of the yarns in the fabric (Hu, 2008).

Like as woven structure, knitting structure has also an important effect on abrasion characteristics of knitted fabrics. Average abrasion resistance values of interlock knitted fabrics are higher than rib and single jersey fabrics (Özgüney et al., 2008). The reason of that is more stabile, thicker and voluminous structure of the interlock fabrics (Akaydın, 2009). Course length for the knitted fabrics is so important that the weight loss percent after abrasion increases with increasing course length. Open, slack knitted fabric structure is abraded more than denser fabrics (Kaloğlu et al., 2003).

The fabric mass per square meter and fabric thickness that are the main structural properties of fabrics have an effect on abrasion resistance. Higher values of these factors ensure higher abrasion resistance.

The other parameter that affects the abrasion is thread density of the fabric. The more threads per unit area in a fabric are the less force to each individual thread is, therefore the fabrics with a tight structure have higher abrasion resistance than those with a loose structure. However as the threads become jammed together they are the unable to deflect under load and thus absorb the distortion (Saville, 1999).

The literature contains papers dealing with the abrasion resistance of the specific type of the fabrics. One of them characterizes certain properties of flocked fabrics produced from different fibre type by measuring the abrasion resistance. Abrasion resistance of the flocked fabrics is related to the flock fibre length and density of flock fibre ends. The flocked fabrics with low flock fibre density and high flock fibre length show more resistance to abrasion in comparison with the flocked fabrics which have high flock fibre density and short flock fibre length. The wet rubbing resistance of the flocked fabrics is less than the dry rubbing resistance (Bilişik, 2009).

Another study is about the performance of upholstery fabrics woven with chenille yarns. Chenille yarn material, yarn twist, and pile length have a significant effect on the abrasion resistance of the chenille yarns and fabrics. Twist levels and pile lengths affect yarn cohesion. There is an improvement in abrasion resistance of the fabrics with increasing twist, pile length, and the use of natural fibres as pile materials, which may be due to increasing frictional behavior between the pile and lock yarns (Özdemir & Çeven, 2004).

3.1.4 Finishing process

Finishing treatments, the types and concentration of the chemicals used in the treatment processes are also the parameters affecting the abrasion characteristics of the fabrics.

Grey fabrics have lower abrasion resistance compared to dyed fabrics with the same construction. During the dyeing operation, fibres on the fabric surface will cling to it, hence the fabric will achieve a closer state, and the movement of fibres within the yarn will be limited (Akaydın 2009, 2010).

Laundering process affects the abrasion resistance. The abrasion resistance of both undyed and dyed fabrics is negatively influenced by the laundering treatment (Candan et al., 2000). The degree of damage in fibres within the fuzz entanglements tends to increase with an increased number of launderings, and that the damage varies from small cracks and fractures to slight flaking depending on the fabric and yarn (Candan & Önal, 2002).

Another process that is important for fabric abrasion is bleaching and enzymatic process. The fabrics applied bleaching and enzymatic processes have higher abrasion resistance with regard to grey knitted fabrics. However as enzymatic treatment is applied to the dyed fabrics the abrasion tendency become worse compared to non-enzymatic dyed fabric (Kretzschmar et al., 2007)

Nano-silicone softener treatment causes decrease in abrasion resistance of the fabrics. The mass loss ratios of the samples with nano-silicone softeners are higher than mass loss ratios of the samples without nano-silicone softener. It is the probable result of fibre mobility inside the fabric which is increased by nano-silicone softener. Silicone softeners provide better wrinkle recovery, tear strength, and abrasion resistance than the cationic softener for 100% cotton woven fabric (Çelik et al., 2010).

The laser fading process is acknowledged as a very strong alternative compared to the conventional physical and chemical processes used for aged-worn look on denim fabrics. Even with the lower pulse time of laser beams, abrasion resistance significantly decreases after fading process and with the higher pulse times, the decrease in abrasion resistance values is much more apparent (Özgüney et al., 2009).

3.2 Methods for testing abrasion resistance of fabrics

Most abrasion tests depend on applying energy to the fabrics and measuring their response to it. The manner of transferring the energy from machine to the fabric is different for different machines, but the basic principles are the same (Abdullah et al., 2006).

There are three types of abrasion in terms of occurrence; flat, edge and flex abrasion. Therefore different abrasion test methods have been described by the abrasion type, the test head movement or testing device setup. The differences among the procedures include the type of equipment, abradant (the material that rubs against the specimen), material used (including woven, nonwoven, and knit apparel fabrics, household fabrics, industrial fabrics, and floor coverings) and assessment method. In all of the test methods, the tested specimen is rubbed in a particular manner against an abradant which may be a fabric, or a emery sheet for either a certain amount of time for a certain number of strokes or cycles (Kadolph, 2007).

ASTM and ISO define several methods to quantify abrasion resistance of textile materials and introduce methods for the evaluation of abraded fabrics. However, there is not a linear relationship between successive measurements using any of these methods and progressive amounts of abrasion (Savilla, 1999). In Table 1, these test methods and relevant test equipments are given.

Test Standard		Testing Device / Method
ASTM D 4966	Standard Test Method for Abrasion Resistance of Textile Fabrics	Martindale Abrasion Tester
ASTM D 3884	Test Method for Abrasion Resistance of Textile Fabrics	Rotary Platform Double-Head (RPDH)
ASTM D 3885	Test Method for Abrasion Resistance of Textile Fabrics	Flexing and Abrasion Method
ASTM D 3886	Test Method for Abrasion Resistance of Textile Fabrics	Inflated Diaphragm
ASTM D 4157	Test Method for Abrasion Resistance of Textile Fabrics	Oscillatory Cylinder Method
ASTM D 4158	Test Method for Abrasion Resistance of Textile Fabrics	Uniform Abrasion Method
AATCC-93 Test Method	Abrasion Resistance of Fabrics: Accelerator Method	Accelerator Method
ISO 12947-1	Determination of the abrasion resistance of fabrics by the Martindale method Part 1: Martindale abrasion testing apparatus	Martindale Abrasion Tester
ISO 12947-2	Determination of the abrasion resistance of fabrics by the Martindale method Part 2: Determination of specimen breakdown	Martindale Abrasion Tester
ISO 12947-3	Determination of the abrasion resistance of fabrics by the Martindale method Part 3: Determination of mass loss	Martindale Abrasion Tester
ISO 12947-4	Determination of the abrasion resistance of fabrics by the Martindale method Part 4: Assessment of appearance change	Martindale Abrasion Tester

Table 1. Test methods of abrasion resistance of fabrics

3.2.1 Flat abrasion

Flat abrasion occurs when flat objects is rubbed to a flat material. Flat resistance tends to be good for most materials because the force of rubbing is distributed over a wide area. However for many products flat abrasion resistance is assumed to occur when the curve is gradual or the bend is shallow such as what occurs in a shirt or jacket as it bends across the back of the wearer or on the seat of an upholstered chair (Kadolph, 2007). Martindale Tester, Taber Abraser, Uniform Abrasion Tester are the instruments working on the flat abrasion mechanism.

In Martindale abrasion resistance tester (Figure 4), circular specimens are abraded under known pressure against a standard fabric. Abrasion resistance is measured by subjecting the specimen to rubbing motion in the form of a geometric figure, denoted Lissojous motion (Figure 5), that is, a straight line, which becomes a gradually widening elipse, until it forms another straight line in the opposite direction and traces the same figure again under known conditions of pressure and abrasive action (ASTM D 4966). The advantage of the Martindale abrasion test is that the fabric sample gets abrasion in all directions. Stress develops along the fiber from the force acting transverse to the fiber axis as a result of surface friction; the magnitude of surface friction developed is directly related to the harshness of standard worsted fabric abradant.

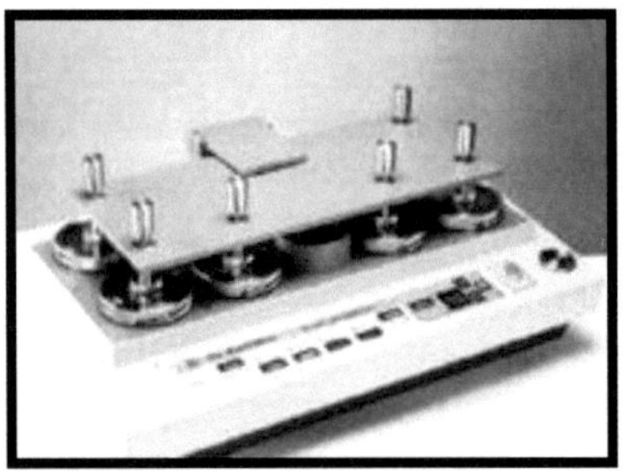

Fig. 4. Martindale Abrasion Tester

Fig. 5. Lissojous motion

Specimens are circular of either 38 mm or 140 mm in diameter. Normally the abradant is silicon carbide paper or woven worsted wool of which the specifications are given in Table 2, mounted over felt. Polyurethane foam disk is placed under the specimen for fabric having a mass/unit area less than 500 g/m^2. The small test specimen is sitting on the large abradant and then cycled backwards and forwards. If assessment of appearance change needs to be carried out, then larger test pieces (140 mm in diameter) are required. The roles are reversed and the abradant is placed in the holder with the specimen as the base platform. A force of either 9 (for apparel fabrics) or 12 kPa (for upholstery and technical fabrics) is applied to the top of the specimen to hold it against the abradant. The standard abradant should be replaced at the start of each test and after 50 000 cycles if the test is to be continued beyond this number (Hu, 2008; Saville, 1999; Özdil, 2003).

	Warp	Weft
Yarn lineer density	R63, tex/2	R74, tex/2
Threads per unit length	17/cm	12/cm
Single twist	540 ±20 tpm 'Z'	500 ±20 tpm 'Z'
Twofold twist	450 ±20 tpm 'S'	350 ±20 tpm 'S'
Fibre diameter	27.5 ±20 μm	29 ±20 μm
Mass per unit area of fabric, min	5.8 oz/yd² (195 g/m²)	

Table 2. Specification for standard wool abrasion fabric (ASTM D 4966)

The rotary platform double head method (Taber abrader) can be used for most fabrics. The specimen is abraded using rotary rubbing action under controlled conditions of pressure and abrasive action, and subjected to multidirectional abrasion using a rotary rubbing action. Abrasive heads of rubber based compound simulate mild and harsh abrasion. The test specimen, mounted on a turntable platform, turns on a vertical axis, against the sliding rotation of two abrading wheels. The fabric is subjected to the wear action by two abrasive wheels pressing onto a rotating sample. The wheels are arranged at diametrically opposite sides of the sample so that they are rotated in the opposite direction by the rotation of the sample. One abrading wheel rubs the specimen outward toward the periphery and the other, inward toward the center. The resulting abrasion marks form a pattern of crossed arcs over an area of approximately 30 cm^2 (Figure 6a). Load adjustment for varying the load of the abrader wheels on the specimen is possible (ASTM D 3884).

In ASTM D 4158, the uniform abrasion testing machine is used. In the apparatus, a specimen is mounted in a holder and abraded uniformly in all directions in the plane and about every point of the surface of the specimen. The Uniform Abrasion Tester (Figure 6b), consists of the abradant mounted at the lower end of a shaft, weights placed on the upper end of the shaft to produce constant pressure between abradant and specimen throughout the test, lever and cam for raising and lowering the abradant, shaft, and weights. Essentially, the surface of the abradant lies in a plane parallel to the surface supporting the specimen and presses upon the specimen. The abradant and specimen rotate in the same direction at very nearly but not quite the same angular velocity (250 rpm) on noncoaxial axes which are parallel to within 0.0025 mm (0.0001 in.). The small difference in speed is to permit each part of the specimen to come in contact with a different part of the abradant at each rotation. Each rotation is equivalent to one cycle (ASTM D 4158).

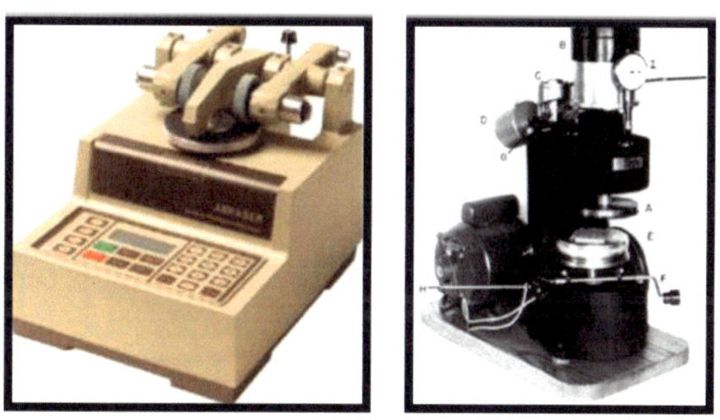

(a) (b)

Fig. 6. (a) Rotary Platform Double Head Abrader, (b) Uniform Abrasion Testing Machine

3.2.2 Flex abrasion

Flex abrasion is the most common abrasion type to which a textile product is subjected. In flex abrasion the material is bent or flexed during rubbing. Flex abrasion occurs in apparel, furnishings and industrial products. Very little of the surface of most products is completely flat during usage, therefore flex abrasion tests may reflect the usage conditions more (Kadolph, 2007). Oscillatory cylinder, inflated diaphgram abrasion tester, flex abrasion testing machine are the instruments working on flex abrasion mechanism.

In ASTM D 4157 oscillatory cylinder method, abrasion resistance is measured by subjecting the specimen to unidirectional rubbing action under known condition of pressure, tension, and abrasive action. During the test fabric is pulled tight in a frame and held stationary. Oscillating Cylinder Section, equipped with edge clamps to permit mounting of a sheet of abrasive material over its surface, capable of oscillating through an arc. Individual test specimens cut from the warp and weft directions are then rubbed back and forth using an ACT approved #10 cotton duck fabric as the abradant. The procedure often is used for upholstery, leather and other materials (ASTM D 4157).

In inflated diaphgram method (Figure 7a), a specimen is abraded by rubbing either unidirectionally or multidirectionally against an abradant having specified surface characteristics. A specimen is held in a fixed position and supported by an inflated rubber diaphragm which is held under constant pressure. The abradant is mounted upon a plate, which is rigidly supported by a double-lever assembly to provide for free movement in a direction perpendicular to the plane of the reciprocating specimen clamp. The abradant plate assembly should be well balanced to maintain a vertical pressure equivalent to a mass of 0 to 2.2 kg (0 to 5 lb) by means of dead weights (ASTM D 3886).

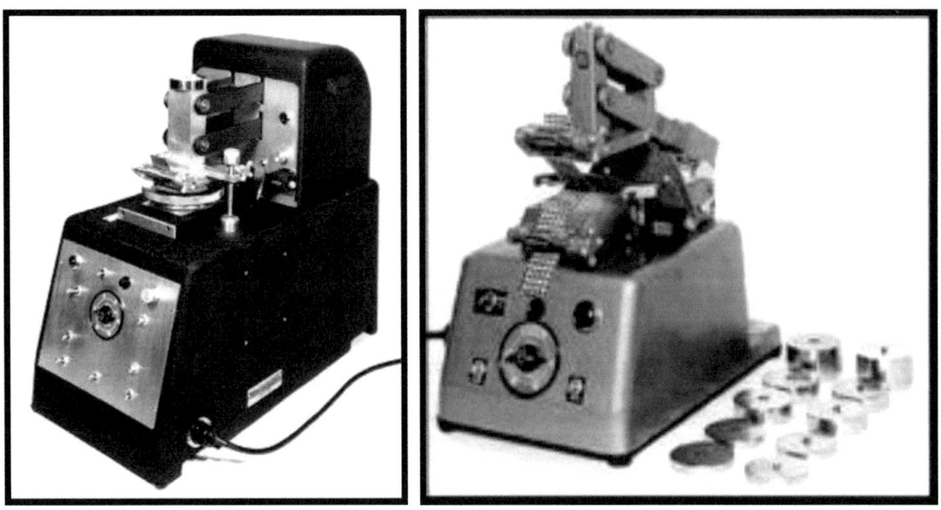

(a) (b)

Fig. 7. (a) Schematic Diagram of Inflated Diaphragm Abrasion Tester, (b) Commercial Flexing and Abrasion Tester

In ASTM D 3885, abrasion resistance is measured by subjecting the specimen to unidirectional reciprocal folding and rubbing over a specific bar under specified conditions of pressure, tension, and abrasive action by Flex Abrasion Testing Machine (Figure 7b). This test method is useful for pretreating material for subsequent testing for strength or barrier performance. The pressure and tension used is varied, depending on the mass and nature of the material and the end-use application. Testing machine consist of Balanced Head and Flex Block Assembly that has two parallel, smooth plates. The balanced head is rigidly supported by a double lever assembly to provide free movement in a direction perpendicular to the plate of the flex block. A positioning device is provided to position the flexing bar and yoke assembly while loading the specimen such that the edge of the flexing bar is parallel to the fold of the specimen during the test (ASTM D 3885).

3.2.3 Edge and Flex abrasion

In edge abrasion the material is folded back on itself while it is being abraded. In products, pleats, folds, cording, cuffs, collars, hems, pocket flaps are most subject to edge abrasion. Most products show damage along edges before damage elsewhere in the product becomes apparent because the force is concentrated on a small portion of the material (Kadolph, 2007). So the edge abrasion is a harsh measure of fabric's resistance to wear.

The accelerator abrasion tester (Figure 8) has an action that is quite different from most other abrasion testers. In the test a free fabric specimen is driven by a rotor inside a circular chamber lined with an abrasive cloth. The specimen suffers abrasion by rubbing against itself as well as the liner. In AATCC 93 method, each specimen is subjected to flexing, rubbing, shock, compression and other mechanical forces while the specimen is tumbling. Abrasion is produced thorough the specimen by rubbing of yarn against yarn, fibre against fibre, surface again surface and surface against abradant (AATCC 93; Kadolph, 2007)

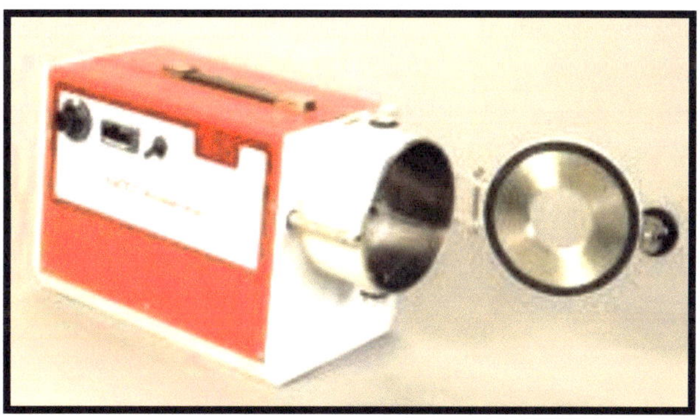

Fig. 8. Accelerator abrasion tester

3.3 Factors affecting abrasion test results

During testing, the resistance to abrasion is also greatly affected by the conditions of the tests, such as the nature of the abradant, variable motion of the abradant over the area of specimen, the tension of the specimen, the pressure between the specimen and the abradant, and the condition of the specimen (wet or dry).

The type of abrasion like flat, flex or edge abrasion or a combination of more than one affect the test results as expected.

In many fabrics the abrasion resistance in the warp direction differs from that of the weft direction. Ideally the rubbing motion used by an abrasion machine should be such as to eliminate directional effects. In jacquard and textured fabrics, the test sample should include both the textured and different parts, sensitive to abrasion.

Abradants can consist of anything that will cause wear. A number of different abradants have been used in abrasion tests including standard fabrics, abrasive paper (glass paper, sand paper etc.) or stones (aluminum oxide or silicon carbide), steel plates and metal 'knives'. The nature of abradants and the type of action control the severity of the test. For the test to correspond with actual wear in use it is desirable that the abrasive should be similar to that encountered in service. The abradant itself wear during the test so they should be replaced after a certain usage time (Hu, 2008 ; Saville, 1999).

Abrasion tests are all subject to variation due to changes in the abradant during specific tests. The abradant must accordingly be discarded at frequent intervals or checked periodically against a standard. With disposable abradants, the abradant is used only once or discarded after limited use. With permanent abradants that use hardened metal or equivalent surfaces, it is assumed that the abradant will not change appreciably in a specific series of tests. Similar abradants used in different laboratories will not change at the same rate, due to differences in usage. Permanent abradants may also change due to pick up of finishing or other material from test fabrics and must accordingly be cleaned at frequent intervals.

The measurement of the relative amount of abrasion may also be affected by the method of evaluation and may be influenced by the judgment of the operator (ASTM D 3884).

The pressure between the abradant and the sample and the test speed affects the severity and rate at which abrasion occurs. It has been shown that using different pressures can seriously alter the ranking of a set of fabrics when using a particular abradant. Excessive pressure and testing speed can result in shorter testing time but however, the results can not simulate the normal usage conditions. The tension of the mounted specimen is also important, and it should be same for all samples. Tension during the test is provided by backing foam or inflated diaphragm.

3.4 Evaluation methods of abrasion tests

There are different options for assessment of the tested fabrics as given below.

- First one is finding endpoint which counts the number of cycles until the fabric ruptures, two or more yarns have broken or a hole appears. The end point is different according to fabric type; for woven structure abrading is continued until two threads are broken while one broken yarn causing a hole for knitted fabric. Removing of the fabric fusses completely, and occurring a hole in 0.5 mm diameter is the end points for fused and for nonwoven fabrics respectively.

- Second option is assessment of the abraded fabric for a set time or number of cycles in terms of some aspect such as loss of mass, loss of strength, change in thickness or other relevant property. The loss of any property before and after abrasion is reported as loss in the relevant property or as a percentage calculated by the formula:

Loss percentage in relevant property (%) = ((A-B)/A)*100, where: A = relevant property before abrasion, and B = after abrasion

According to ISO 12947-3 standard, the weight loss in certain cycles is determined based on the end breaking cycle, then at the every intervals the abraded sample is taken from the device, weighted in the sensitivity of 1 mg, percentage of the weight loss is calculated and the graph is plotted.

Series of experimental	Abrasion cycles at the breakage of the sample	Abrasion cycles for determination of weight loss
a	≤1000	100, 250, 500, 1000, (1250)
b	> 1000 ≤ 5000	500, 750, 1000, 2500, 5000, (7500)
c	> 5000 ≤ 10000	1000, 2500, 5000, 7500, 10000, (15000)
d	> 1000 ≤ 25000	5000, 7500, 10000, 15000, 25000, (40000)
e	> 25000 ≤ 50000	10000, 15000, 25000, 40000, 50000, (75000)
f	> 50000 ≤ 100000	10000, 25000, 50000, 75000, 100000, (125000)
g	> 100000	25000, 50000, 75000, 100000, (125000)

Table 3. Intervals for the measurement of weight loss

- The third method is the evaluation of any visual changes that occurred as a result of abrasion test. The end point is reached when there is a change in shade or appearance that is sufficient to cause a customer to complain or the visual appearance after the predetermined abrasion cycles (ISO 12947-4). Visual change considers the effect that a specified number of cycles have on the lustre, colour, surface nap or pile, pilling, matting or any chance resulting from abrasion. Average number of rubs required to reach a gray scale rating of three or lower is recorded while assessing colour related changes.

None of the above assessment methods produces results that show a linear or direct comparison with one another (Bird, 1984).

4. Yarn abrasion

The issue of yarn abrasion has been defined both as the attrition of yarn upon itself (yarn to yarn) or another surface (yarn to wire, etc.) and the resistance of yarn to suffer damage when abraded upon a surface (Johns, 2001). A yarn which is being knitted or woven into a fabric or wound onto a package runs around many guides during the process. As a yarn passes over a metallic surface, either it gets abraded or it abrades the metallic surface, which is an undesirable effect causing wear of machine parts (Saville, 1999). Yarn abrasion resistance is an important factor affecting weavability (Lawrence, 2003). The yarn-to-yarn friction between warp and weft yarns at every crossover is important in determining the shear property and consequently the formability of woven fabrics (Liu et al., 2006).

The factors, which affect the yarn abrasion, can be detailed as; yarn friction, yarn type, twist multiple, hairiness, contaminants and applied lubricants (Johns, 2001, as cited in Alterman, 1985; Barella, 1989; Beckert, 1999; Chattopadhya& Banerjee, 1996; Steve, 1986, Süpüren et al., 2008). Average surface length which is highly correlated with the abrasion resistance of yarns is also introduced as a new yarn structural parameter. The length of fibres exposed on the yarn surface depends on their interactions with surrounding fibres on the yarn. The yam abrasion resistance was found inversely proportional to the average surface length (Choi & Kim, 2004). For synthetic yarns, it is indicated that, abrasion behavior of the yarns is associated with the fundamental properties such as the polymer and fibre molecular weight, undrawn fibre orientation and crystalline structure, drawn fibre properties, and elevated temperature post-treatment (Ross & Wolf, 1966).

The coefficient of friction is a significant characteristic of textile yarns, because it defines the amount of resistive shear force a yarn will exert and have exerted upon it during fabric forming or preparation processes (Thomas & Zeiba, 2000). As the friction between the yarn and the abradant increases, the abrasion resistance decreases (Johns, 2001, as cited in Jeddi & Sheikhzadeh, 1994).

Yarn type is another important parameter for the abrasion resistance. Generally, ring-spun yarns have higher snagging tendency and lower abrasion than rotor yarns (Lawrence, 2003). Air-jet spun yarns are less resistant to abrasion than ring-spun yarns, because the former are bound by outer wrapper fibres only and have no internal twist (Thomas & Zeiba, 2000). The abrasion resistance of the sirospun yarn is better than that of the two-plied and the single yarns (Sun & Cheng, 2000).

Resistance to abrasion increases as the twist multiplier increases (Johns, 2001). At low twist levels, yarns can be easily splitted therefore yarns with high twist levels are abraded less than yarns with low twist levels.

The abrasion resistance of the two-plied yarn depends on both single-yarn twist and ply twist. Single-yarn twist and ply twist have a more influential effect on the yarn-to-yarn and yarn-to-pin abrasion resistances respectively of cotton two-ply yarns (Palaniswamy & Mohamed, 2006).

It is known that during weaving, the yarns experience the abrasive action of the moving loom parts such as heddle eyes, reeds, whip rolls, and picking elements (Goswami et al., 2004). Sizing is applied to the warp yarns to improve their abrasion resistance and strength and to reduce friction (Lawrence, 2003). The yarn abrasion resistance was found higher in sized yarn with a higher twist level (Kovačević & Gordoš, 2009). The size formulations used for spun yarns (including blends) also contain other ingredients such as lubricants and binders. The lubricants help to reduce the friction and abrasion between the adjacent yarns and between yarns and heddles, dropwires, shuttles, rapiers, or projectiles (Goswami et al., 2004).

4.1 Measurement methods of testing yarn abrasion

There are various measurement methods to measure abrasion properties of yarns, including yarn on yarn abrasion and yarn external abrasion, which comprises different test procedures.

4.1.1 Yarn on yarn abrasion

The yarn-on-yarn abrasion test is described in ASTM D 6611, "Standard Test Method for Wet and Dry Yarn-on-Yarn Abrasion Resistance". A length of yarn is inter-wrapped in contact with itself between three pulleys that are positioned in a defined geometry to produce a specific intersection angle. A weight is attached to one end of the yarn to apply a prescribed tension. The other end is drawn back and forth through a defined stroke at a defined speed until the yarn fails due to abrasion upon itself within the inter-wrapped region (Figure 9). The yarn abrasion test can be conducted in either the dry state or the wet state (ASTM D 6611).

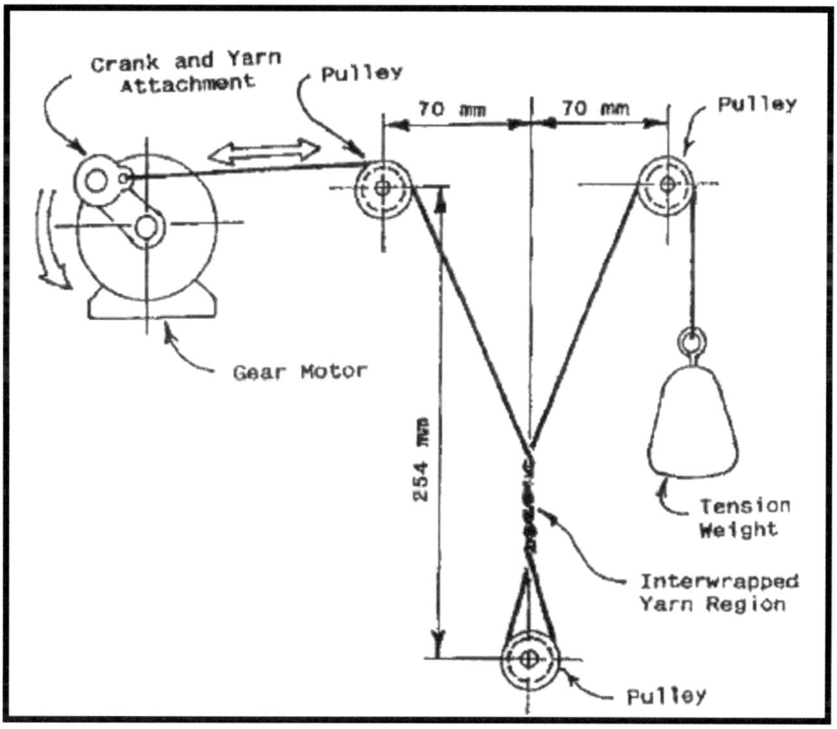

Fig. 9. Yarn-on-yarn abrasion setup (ASTM D 6611)

Zweigle Staff-Tester G 556 (Figure 10), measures accurately abrasion and breakage tendency before further processing. The yarn runs with constant tension via the hysteresis brake into the tester. The thread is guided over two easy-going rollers in such a way, that the running-in section coils around the running-out section over an angle of approximately 180 degrees. At this coiling point there is intense thread-to-thread friction. Various metal or ceramic elements can be placed into the running path of the thread. A funnel below the coiling point collects the parts rubbed-off from the yarn or avivage. The built-in blast exhausts the dust into a small, easily removable filter. By using an accurate balance the weight of the yarn and/or avivage rubbings can be determined (http://citeseerx.ist.psu.edu).

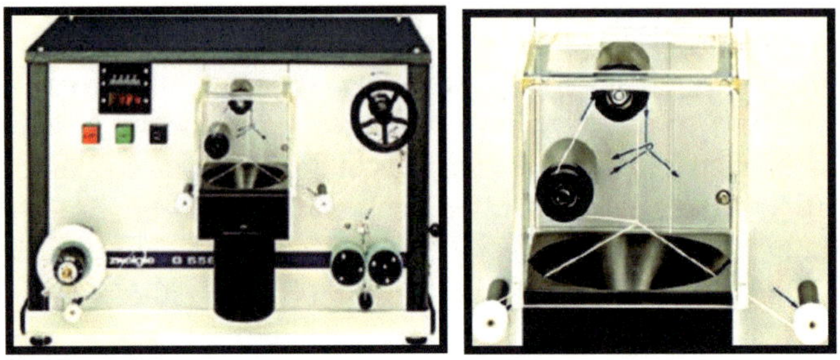

Fig. 10. Staff tester (http://citeseerx.ist.psu.edu)

Lint generation (lint shedding) test can be conducted by Lawson-Hemphill's CTT (Constant Tension Transport) instrument to measure yarn to yarn abrasion (Figure 11). The instrument is a dynamic yarn transport system with the ability to maintain a specific yarn tension and let the yarn run at any selected speed up to 360 m/min. It simulates the production conditions in the testing laboratory and therefore it gives idea about the production performance (Thavamani, 2003).

The CTT Lint Generation Test determines how much lint a yarn will generate under certain manufacturing conditions, including: yarn-to-yarn thread paths, yarn-to-metal thread path, yarn-to-ceramic and yarn-to-needles, sinkers and reeds (at any angle). During the lint test, the yarn is wrapped around itself. As the yarn is moving, the generated lint will be collected on a piece of paper under the vacuum sealed enclosure. The amount of lint that is generated is expressed as mg/1 km. Therefore, abrasion properties of different yarns can be compared based on their weight loss (www.lawsonhemphill.com).

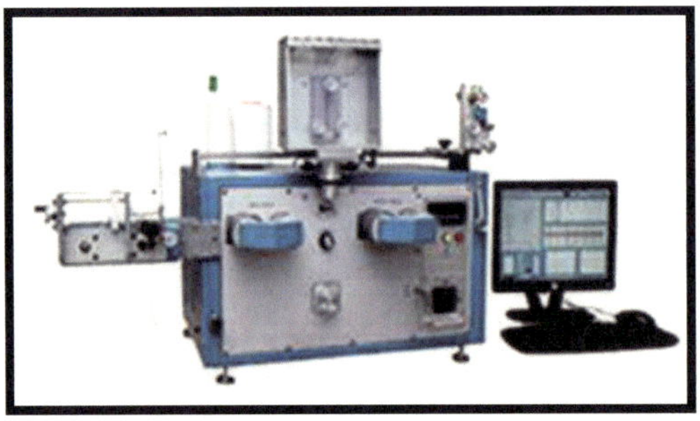

Fig. 11. CTT Instrument with Lint Generation Test Module (www.lawsonhemphill.com)

4.1.2 Yarn external abrasion

External abrasion resistance is important in many applications and there are different measurement methods to evaluate yarn abrasion resistance.

Oscillating stress tester (Figure 12a) incorporates the abrasion element consisting of three rows of case-hardened steel pegs. Yarns hanging from the jaws just touch the pegs in the top and bottom rows, and the middle pegs can be traversed sideways to deflect the yarns around by a desired amount. In a test, the yarn hanging from the jaws passes to one side of the corresponding top and bottom pegs and around the opposite sides of the displaced middle peg, and thus abrade the yarn when reciprocated. Owen and Locke also modified the instrument to apply abrasion by the rotating steel shaft, as shown in Figure 12b. The shaft was rotated by a belt driven from a small electromotor to provide abrasion to yarns at the point of contact (Goswami et al., 2004).

(a) (b)

Fig. 12. Oscillating stresses abrasion tester (Goswami et al., 2004)

Using these modified equipment, Owen and Locke subjected yarns to a fixed number of abrasion cycles and measured the breaking strength of such abraded yarns. The results were expressed in terms of percent deterioration in the breaking load, calculated as follows:

$$\text{deterioration } (\%) = \frac{MSYBA - MSYAA}{MSYBA} \times 100$$

where, $MSYBA$= mean strength of yarn before abrasion and $MSYAA$= mean strength of yarn after abrasion (Goswami et al., 2004).

As a yarn passes over a metallic surface (Figure 13), either it gets abraded or it abrades the metallic surface, which is an undesirable effect causing wear of machine parts (Johns, 2001). Therefore, the abrasiveness of the yarns is another abrasion characteristic of the yarns and can be tested by Lawson-Hemphill CTT (Constant Tension Transport) tester. In this test, the wires used in testing are secured at one end, stretched over a pulley and secured at the other end to a lever. When the wire is cut, the weight hanging on the lever enables the activation of auto stop mechanism. Suitable weight has to be used on the lever to give support to the wire to keep it straight when the wire is passing over it without stretching it unnecessarily (Thavamani, 2003). The yarn is run over a tensioned standardized copper wire and the abrasion factor is measured as the length of yarn it takes to cut the wire or, the surface destruction is measured with a microscope, after running of the yarn in a fixed length (Das & Hati, http://www.lawsonhemphill.com/pdf/lawson-hemphill-news-006.pdf)

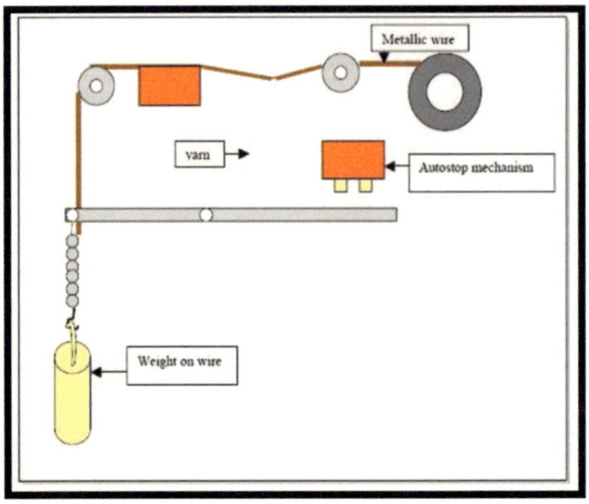

Fig. 13. Diagram to show the CTT set up of yarn and metallic wire (Thavamani, 2003)

Another instrument is Shirley yarn abrasion tester (Figure 14a), which consists of two reciprocating bars: one is made of hardened steel and the other is covered with the standard abradant used in the Martindale fabric abrasion tester. Eight yarn specimens are tested simultaneously. Yarns are threaded from the fixed holders and clipped onto the flexible holders where sensors are attached. The initial tension exerted on each yam is 0.5N. When a yam breaks, the flexible holder falls, a signal is sent to the control unit, and the number of rubs for that particular yarn is recorded (Choi & Kim, 2004).

The effect of yarn count and material type on the abrasiveness property is also significant for fancy yarns. Wool fancy yarns are more abrasive than the acrylic yarns and thicker yarns are more abrasive than the thinner yarns as well (Süpüren et al., 2008).

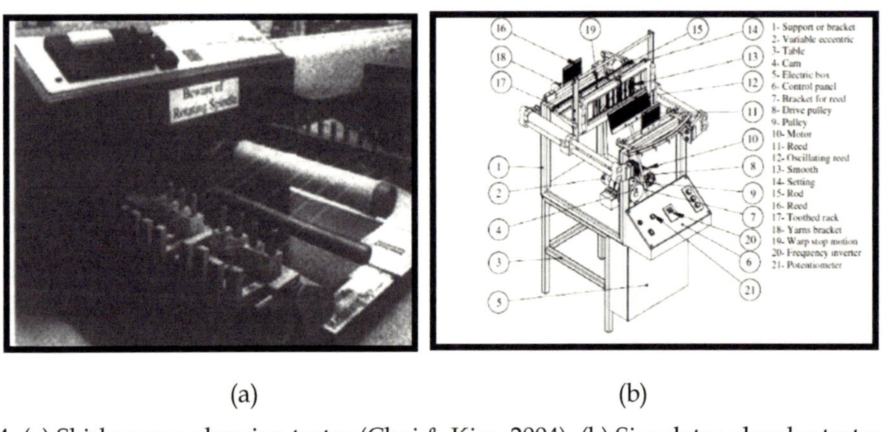

(a) (b)

Fig. 14. (a) Shirley yarn abrasion tester (Choi & Kim, 2004), (b) Simulator abrader tester (Boubaker et al., 2011)

A new abrader tester (Figure 14b) which simulates the weaving process, combines forces of traction, bending, yarn-to-yarn and yarn-to-metal friction. The repetitive deformations of spliced and parent yarns simulate the abrasion action on the weaving machine (Boubaker et al., 2011).

5. Abrasion resistance of socks

Socks, which are a necessary item of clothing, need to be comfortable, affordable and retain their quality throughout their life. The most significant problem is abrasion which can greatly reduce the materials life. Abrasion usually occurs on the heel, sole and toes of the socks. The span life of the socks is shorter than other textile materials because of higher abrasion within the shoes, slippers or even the ground during usage. The first stage of abrasion unrevals of the loose fibres from the fabric surface, eventually breakdown of fibres and a hole occur. If the sock consists of synthetic fibres with natural fibres, during rubbing action natural fibres, which give the desirable properties of the sock, move away, only synthetic fibres remain. This situation stated as thinning and gives the sock undesirable appearance and decreases the overall fabric thickness (Figure 15).

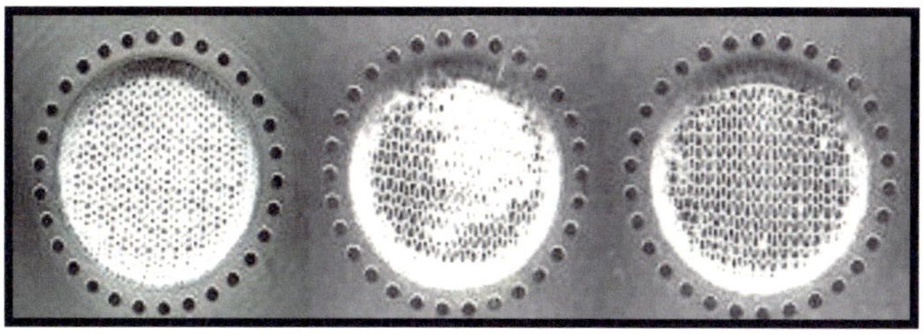

(a)

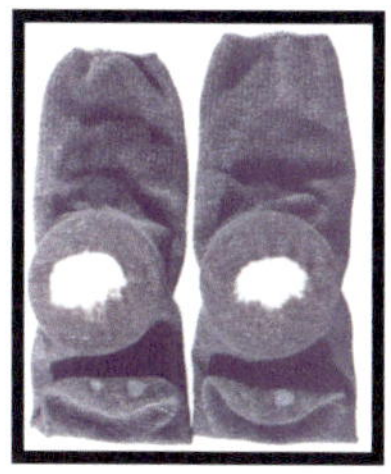

(b)

Fig. 15. (a) The thinning appearance of a Co/PA sock at the heel during abrasion resistance tests, (b) abraded socks (Özdil et al., 2005, 2009)

The Sock Testing Consortium accepted three methods for abrasion resistance of socks. These are CSI Stoll method, ILE SCR method and in present widely used Martindale method (Özdil et al., 2005). A modified specimen holder for the Martindale abrasion tester, which stretches the knitted material, is used for socks' abrasion (Figure 16). The holder takes a standard size 38mm diameter sample which is held to size by a pinned ring. A flattened rubber ball is pushed through the sample as the holder is tightened thus stretching it, test is carried out as fabric tests. The sample is inspected at suitable intervals until a hole appears or the material develops an unacceptable level of thinning (Özdil et al., 2009).

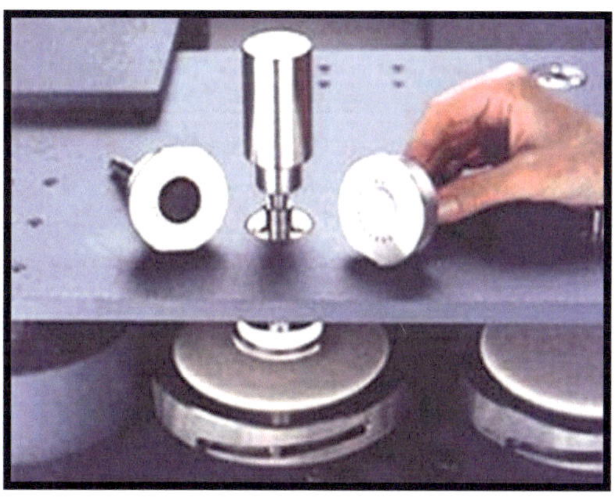

Fig. 16. Setting of sock kit on Martindale apparatus

There is not too much research related with sock abrasion. In one of the research abrasion resistance of 7 different types of socks consisting of different rates of Co-PA were searched with both laboratory tests and usage tests (Özdil et al., 2009, as cited in Wisniak & Krzeminska, 1987). Pilling test device was used for abrasion resistance and abrasion time of the sample assessed. It was found that the results from the laboratory and the usage tests were different. In another study the abrasion resistance of terry socks was investigated (Özdil et al., 2009, as cited in Miajewska & Kazmierczak, 1983). PA yarns for ground, wool and wool blends (wool+ PA, PAC+PA, wool+PAC) also cotton and cotton blends (Co+PA, Co+viscone+PA) for pile were used. It was found that the result of the wool and wool blends are evenly matched to each other and abrasion resistance of the cotton socks was better than wool socks. They used different yarns for piles to increase the abrasion resistance of the socks and found that the yarns spun with wool -PES blends gave the best results. The results of the PAC–wool yarns and PAC-linen yarns followed it respectively. Increased yarn twist, adding PA and folded PAC yarns increases abrasion resistance of the socks are the other results of them.

The effects of the yarn parameters and some finishing process on the abrasion resistance of socks were researched in detail (Özdil et al., 2009). It was found that the abrasion resistance value of socks can be increased by a number of measures; usage of thicker yarns, adding PA and elastic yarns to the structure, increasing the PES ratio in Co-PES yarns. The resistance of wool socks is higher than acrylic ones and the wool ratio in wool-acrylic samples has a positive influence on abrasion resistance. The abrasion resistance increases with mercerization process and decreases with the use of silicone softeners (Özdil et al., 2009).

6. Abrasion resistance in technical textiles

Abrasion resistance is one of the most important properties in special technical textile products. In order to measure the abrasion resistance of these products such as protective clothes, military fabrics, gloves, laminated fabrics, multi layered structures, special measurement procedures were developed.

Car seat is one of the most important parts of the interior of a car and abrasion resistance of the seat fabric is an important parameter for the usage. It is composed of metal structure, filling (molded polyurethane foam) and seat cover including exterior fabric, foam and support material (reinforcement material).The foam has influence on the abrasion resistance of car seats upholstery; increased height and weight of the foam, causes less weight and thickness loss. Non-usage of foam reduced the abrasion resistance significantly (Jerkovic et al., 2010). Fabric type has an important effect on abrasion resistance of car seat covers. The warp knit double bar raschel was found more resistant to abrasion than flat woven, circular knitted flat and warp knit flat fabrics (Pamuk & Çeken, 2008).

An abrasion test for geo-textiles uses a reciprocal back-and-forth rubbing motion of a sandpaper abradant against the geo-textile. The instrument used in this test is a Stoll Flex abrader modified to allow the fabric to be mounted on a stationary platform and the abrading medium on a reciprocating platform. The abrader can be loaded to provide a constant pressure and has an adjustable speed drive to allow the abrasive action to be controlled as required (Hodge, 1987).

In day-to-day applications, abrasion resistance is one of the most important properties that determine the duration of the useful life of the nonwoven articles. In particular, in many industrial filtration applications, filter media can undergo severe abrasive loading. Abrasive damage can manifest itself in two ways. Firstly, the filtration area can be damaged due to repeat contact with hard and sharp particulate materials in the fluid, and can result in pinhole damage over time. This in turn weakens the body of filter media and reduces its retention efficiency, resulting in rapid solids loss. In such cases, the process would have to be stopped and the cloths either changed or occasionally repaired with patches. A second source of abrasive damage is due to attrition between the cloth and the filter machinery. The abrasion resistance of the thermally bonded nonwoven articles is significantly dependent on not only the choice of fibre, but also the construction of the fabric (Wang et.al, 2007; as cited in Ramkumar et al, 2001), which is further influenced by both the web forming and thermal bonding processes (Wang et.al, 2007).

Abrasion has a potential mode of failure for either latex or non-latex medical glove materials. Resistance to abrasion is necessary to maintain barrier integrity during routine tasks such as twisting a capped needle onto a syringe or turning a knurled knob on a piece of equipment. An alternative method (Figure 17) for measuring durability of both latex and non-latex medical glove materials, utilizing abrasion resistance testing, has been developed.

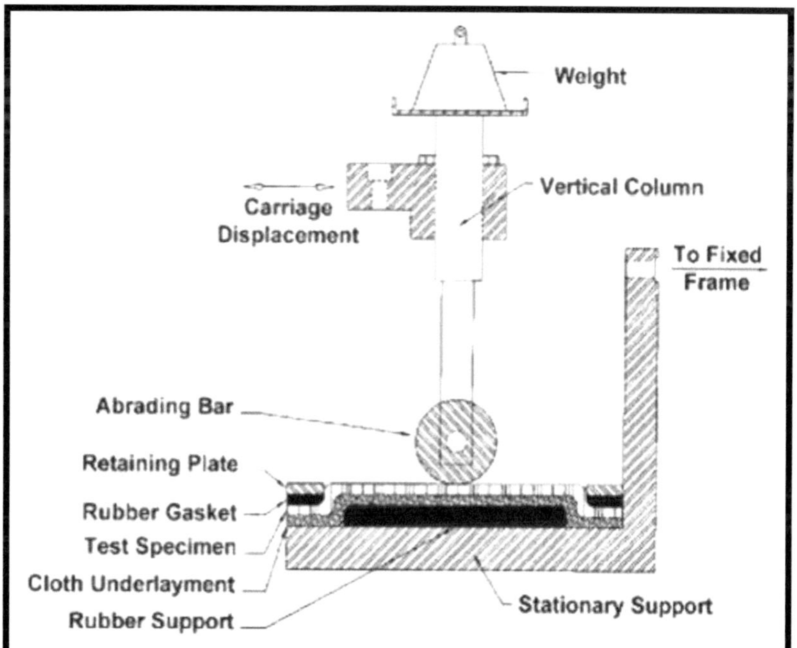

Fig. 17. Glove abrasion resistance apparatus (Walsh et al., 2004)

This device converts the rotary motion of an adjustable-speed, DC gear motor into oscillating linear displacement of a carriage that serves as a translating support for a vertical loading column. The abrader is linearly stroked against a test specimen at a rate that is preset by adjustment of a motor speed control. Along with displaying the preset number of cycles for a particular test sequence, the system control unit also indicates real-time values of test cycles and cycle rate (Walsh et al., 2004).

Established test machines such as the Martindale, revolving drum and Taber abrader were considered for the evaluation of "Protective clothing for motorcycle riders", but they are not capable of testing the multitudinous single and combination materials and constructions present in motorcyclists' protective clothing. The Darmstadt tester and the Cambridge tester are used for this measurement. The Darmstadt machine consists of a `doughnut' of concrete, in the centre of which is situated an electric motor. In testing, the motor is run up to a specified speed at which point the specimen holders are released and freefall the short distance down the axis of the shaft into contact with the surface of the concrete. The specimens continue to spin freely around the drive shaft and in contact with the concrete until coming to a halt. The mass of test specimens both before and after testing is recorded, and the difference established. In Cambridge tester, the test specimen is mounted on a holder which is attached to the free end of a horizontal, rigid pendulum which pivots at the opposite end to the specimen holder. In testing, the pendulum is released and falls onto the moving belt. A fine copper wire of 0.14mm diameter, located across the outer face of the specimen, is cut upon contact with the moving belt and this starts an electronic timer. A second wire is exposed and cut when the specimen is abraded through, which stops the timer and records the time taken to perforate the specimen. The more prolonged the period between contact and perforation, the better-performing the material. Para-aramid laminated constructions for motorcyclists' clothing was found better than polyester laminated construction, layered construction textile, air textured nylon and leather in terms of the results of impact abrasion resistance (Scott, 2005).

7. Conclusion

In this chapter, abrasion which affects serviceability of textile materials was explained in detail. The abrasion mechanism of textiles is a complex phenomenon and associated with the properties of fibers, yarns, fabric structure and applied treatments. Abrasion in textiles such as fabrics, yarns, socks and technical textiles can be measured by different methods. Due to the technological improvements and growing demands on the properties of textile materials, it seems the development of new test techniques and equipments will continue on this issue.

8. References

AATCC Test Method 93 Abrasion Resistance of Fabrics: Accelerator Method

Abdullah, I., Blackburn, R.S., Russell, S.J., Taylor, J. (2006). Abrasion Phenomena in Twill Tencel Fabric, *Journal of Applied Polymer Science,* Vol. 102, pp.1391–1398

Akaydin M. (2009). Characteristics of Fabrics Knitted With Basic Knitting Structures From Combed Ring And Compact Yarn, *Indian Journal of Fibre & Textile Research,* 34, pp. 26-30.

Akaydın M. (2010). Pilling Performance and Abrasion Characteristics of Selected Basic Weft Knitted Fabrics, *Fibres & Textiles in Eastern Europe*, Vol. 18, No. 2 (79)

ASTM D 3884-09 Test Method for Abrasion Resistance of Textile Fabrics (Rotary Platform, Double-Head Method)

ASTM D 3885-11 Test Method for Abrasion Resistance of Textile Fabrics (Flexing and Abrasion Method)

ASTM D 3886-11 Test Method for Abrasion Resistance of Textile Fabrics (Inflated Diaphragm Method)

ASTM D 4157-10 Test Method for Abrasion Resistance of Textile Fabrics (Oscillatory Cylinder Method)

ASTM D 4158-08 Test Method for Abrasion Resistance of Textile Fabrics (Uniform Abrasion Method)

ASTM D 4966-10 Standard Test Method for Abrasion Resistance of Textile Fabrics (Martindale Abrasion Tester Method)

ASTM D 6611-07, Standard Test Method for Wet and Dry Yarn-on-Yarn Abrasion Resistance

Bilişik K. (2009). Abrasion Properties of Upholstery Flocked Fabrics, *Textile Research Journal*, Vol 79 (17), pp. 1625–1632

Bird S L, (1984). *A Review of the Prediction of Textile Wear Performance with Specific Reference to Abrasion*, SAWTRI Special Publication, Port Elizabeth.

Boubaker, J., Hassen, M.B., Sakli, F. (2011). Abrasion Evaluation of Spliced and Parent Yarns with a New Simulator Abrader Tester, *The Journal of Textile Institute*, pp. 1–6

Brown, R. (2006). *Physical Testing of Rubber*, Springer, ISBN 10 0-387-28286-6, Unites States of America

Candan, C, Nergis, U.B., Iridag, Y. (2000). Performance of Open-end and Ring Spun Yarns in Weft Knitted Fabrics, *Textile Research Journal*, Vol. 70. No 2, pp. 177-181.

Candan, C. Önal, L. (2002). Dimensional Pilling and Abrasion Properties of Weft Knits Made from Open-End and Ring Spun Yarns, *Textile Research Journal*, Vol. 72. No 2, pp. 164-169.

Çelik, N., Değirmenci, Z., Kaynak, H.K. (2010). Effect Of Nano-Sılıcone Softener On Abrasıon And Pilling Resistance And Color Fastness Of Knıtted Fabrıcs, *Tekstil ve Konfeksiyon* Vol. 1, pp. 41-47

Choi, K.F., Kim, K.L. (2004). Fibre Segment Length Distribution on the Yarn Surface in Relation to Yarn Abrasion Resistance, *Textile Research Journal* 74, pp. 603-606

Collier, B. J., Epps, H. H., (1999). *Textile Testing and Analysis*, Prentice Hall, New Jersey.

Das, B.R., Hati, S., New Generation Tensile Tester: CTT, http://www.lawsonhemphill.com/pdf/lawson-hemphill-news-006.pdf

Goswami, B.C., Anandjiwala, R.D., Hall, D.M. (2004). *Textile Sizing*, Marcel Dekker, INC., New York Basel, ISBN: 0-8247-5053-5

Hearle, J.W.S., Morton, W.E. (2008). *Physical Properties of Textile Fibres* (Fourth edition), Woodhead Publishing Series in Textiles No. 68, ISBN-13: 978 1 84569 220 9

Hodge, J. (1987). *Durability Testing*, Geotextile Testing and the Design Engineer.ASTMSTP 952, i. E.Fluet, Jr., Ed., American Society for Testing and Materials, Philadelphia, pp. 119-121.

http://citeseerx.ist.psu.edu/viewdoc/download;jsessionid=9D6EE9FE39FA026083B648EFB29BA8C4?doi=10.1.1.113.2764&rep=rep1&type=pdf

http://en.wikipedia.org/wiki/Wear

http://www.lawsonhemphill.com/LH-405-lint-generation-tester.html

Hu, J. (2008). *Fabric testing*, Woodhead Publishing Series in Textiles: Number 76

ISO 12947-3 1998-Determination of the abrasion resistance of fabrics by the Martindale method Part 3: Determination of mass loss

ISO 12947-4 1998-Determination of the abrasion resistance of fabrics by the Martindale method Part 4: Assesment of appearance change

Jerkovic, I., Pallares, J.M., Capdevila, X. (2010). Study of the Abrasion Resistance in the Upholstery Of Automobile Seats *AUTEX Research Journal*, Vol. 10, No1

Johns., J., (2001). *Abrasion Characteristics of Ring-Spun and Open-End Yarns*, North Carolina State University, Master Thesis

Kadolph, S.J. (2007). *Quality Assurance for Textiles and Apparel*, ISBN:156367-144-1, Fairchild Publication

Kaloğlu, F., Önder, E., Özipek, B. (2003). Influence Of Varying Structural Parameters On Abrasion Characteristics of 50/50 Wool/Polyester Blended Fabrics, *Textile Research Journal*, Vol. 73, No. 11, pp. 980-984.

Kaynak, H.K., Topalbekiroğlu, M. (2008). Influence of Fabric Pattern on the Abrasion Resistance Property of Woven Fabrics, *Fibres & Textiles in Eastern Europe*, Vol. 16, No. 1 (66), pp. 54-56

Kovačević, S., Gordoš, D. (2009). Impact of the Level of Yarn Twist on Sized Yarn Properties, *Fibres & Textiles in Eastern Europe*, Vol. 17, No. 6 (77)

Kretzschmar, S.D., Özgüney, A.T., Özçelik, G., Özerdem, A. (2007). Yarns Before and After the Dyeing Process The Comparison of Cotton Knitted Fabric Properties Made of Compact and Conventional Ring, *Textile Research Journal*, Vol. 77, pp. 233-241

Lawrence, C.A. (2003). *Fundamentals of Spun Yarn Technology*, CRC Press Boca Raton London New York Washington, D.C., ISBN 1-56676-821-7

Liu, L., Chen, J., Zhu, B., Yu, T.X., Tao, X.M, Cao, J. (2006). The Yarn-to-Yarn Friction of Woven Fabrics. *In: Proceeding of 9th International ESAFORM Conference on Materials Forming*, UK, April 26-28

Manich, A.M., Castellar, M.D.D., Sauri, R.M., Miguel, R.A., Barella, A. (2001). Abrasion Kinetics of Wool and Blended Fabric, *Textile Research Journal*, Vol.71, pp. 469-474.

Mehta, P.V. (1992). *An Introduction to Quality Control for the Apparel Industry*, ASQC Quality Press

Omeroglu S., Ulku Ş. (2007). An Investigation about Tensile Strength, Pilling and Abrasion Properties of Woven Fabrics Made from Conventional and Compact Ring-Spun Yarns, *Fibres & Textiles in Eastern Europe*, Vol.15(1), pp. 39-42

Örtlek, H.G., Yolaçan, G., Bilget, Ö., Bilgin, S. (2010). Effects of Enzymatıc Treatment on The Performance of Knıtted Fabrıcs Made From Dıfferent Yarn Types, *Tekstil ve Konfeksiyon*,Vol.2 pp. 115-119

Özdemir Ö., Çeven E. K. (2004). Influence of Chenille Yarn Manufacturing Parameters on Yarn and Upholstery Fabric Abrasion Resistance, *Textile Research Journal*, Vol. 74(6), pp.515-522

Özdil, N. (2003). *Kumaşlarda Fiziksel Kalite Kontrol Yöntemleri*, Ege Universitesi, Tekstil ve Konfeksiyon Araştırma Uygulama Merkezi Yayını, No:21, ISBN: 975-483-579-9

Özdil, N., Marmarali, A., Oglakcioglu, N. (2005). A Research on Abrasion Resistance of the Socks, Ege University Scientific Research Project

Özdil, N., Marmarali, A., Oğlakcioğlu, N. (2009). The Abrasion Resistance of Socks, *International Journal of Clothing Science and Technology*, Vol: 21, No: 1, pp. 56-63

Özgüney, A.T., Kretzschmar, S.D., Özçelik, G., Özerdem, A. (2008). , The Comparison of Cotton Knitted Fabric Properties Made of Compact and Conventional Ring Yarns Before and After the Printing Process, *Textile Research Journal*, Vol. 78, pp. 138-147

Özgüney, A.T., Özçelik, G., Özkaya, K. (2009). A Study on Specifying The Effect of Laser Fading Process On The Colour And Mechanical Properties of The Denim Fabrics, *Tekstil ve Konfeksiyon*, Vol.2, pp. 133-138

Paek, S. L. (1989). Pilling, Abrasion and Tensile Properties of Fabrics from Open-End and Ring Spun Yarns1, *Textile Research Journal*, Vol. 59 (10), pp. 577-583

Palaniswamy, K., Mohamed, P. (2006). Effect Of The Single-Yarn Twist And Ply To Single-Yarn Twist Ratio On The Hairiness And Abrasion Resistance Of Cotton Two-Ply Yarn, *AUTEX Research Journal*, Vol. 6, No 2, AUTEX, http://www.autexrj.org/No2-2006/0159.pdf 59

Pamuk G., Çeken F. (2008). Comparatıve Study of the Abrasion Resistance of Automobile Seat Covers, *Fibres & Textiles in Eastern Europe*,Vol. 16, No. 4 (69), pp. 57-61

Ross, S.E., Wolf, H.W. (1966). Abrasion Characteristics of Polypropylene Yarns, *Journal of Applied Polymer Science*, Vol.10, Issue 10, pp.1557 1572

Saville, B.P., (1999). Physical Testing of Textiles, CRC, Woodhead Publishing Limited, Cambridge, England

Scott, R.A. (2005). *Textiles for Protection*, The Textile Institute, Woodhead Publishing Limited, Cambridge, England, pp. 720-722, IBSN: 1-85573-921-6

Sun, M.N., Cheng, K.P.S. (2000). Structure and Properties of Cotton Sirospun® Yarn, *Textile Research Journal*, Vol. 70, pp. 261-268

Süpüren, G., Özdil, N., Ozcelik, G., Turay, A. (2009), Abrasion Characteristics of Various Types of Fancy Yarns, 6th international Conference of Textile Research Division NRC, Cairo, Egypt, 5-7

Thavamani, A. (2003). Interaction of Yarn with Metallic Surfaces, Graduate Faculty of North Carolina State University, Master Thesis

Thomas, H.L., Zeiba, J.M. (2000). Size Lubrication Methods for Air-Jet-Spun and Ring-Spun Warp Yarns, *The Journal of Cotton Science*, Vol. 4, pp. 112-123

Walsh, D.L., Schwerin, M.R., Kisielewski, R.W., Kotz, R.M., Chaput, M.P., Varney, G.W., To, T.M. (2004). Abrasion Resistance of Medical Glove Materials. *Journal of Biomedical Materials Research* - Part B Applied Biomaterials, Vol.68 (1), pp. 81-87.

Wang, X.Y., Gong, R.H., Dong, Z., Porat, I. (2007). Abrasion Resistance Of Thermally Bonded 3D Nonwoven Fabrics, Wear , Vol. 262, pp. 424–431

8

Rubber Abrasion Resistance

Wanvimon Arayapranee
Rangsit University
Thailand

1. Introduction

Abrasion resistance is the ability of a material to resist mechanical action such as rubbing, scraping, or erosion that tends progressively to remove material from its surface. When a product has abrasion resistance, it will resist erosion caused by scraping, rubbing, and other types of mechanical wear. This allows the material to retain its integrity and hold its form. This can be important when the form of a material is critical to its function, as seen when moving parts are carefully machined for maximum efficiency. Abrasion resistant materials can be used for both moving and fixed parts in settings where wearing could become an issue.

The substances usually called "rubber" immediately brings to mind materials that are highly flexible and will snap back to their original shape after being stretched. In fact, there are three structural requirements for a given substance to be a rubber (i) rubber is made up of a polymer chain, liner or branched; (ii) the chain is flexible; and (iii) the chain is longer than a certain threshold length. Because the rubber is compliant and tough, it can easily absorb and survive a single strike of large deformation. However when used in contact with moving parts, a process of micro-tearing can occur on the rubber surface around the sharp asperities, gradually removing the material and finally terminating the functional life of the rubber. In many applications, abrasive wear is the major failure mode of rubbers. In normal materials, a rough surface is made smooth by repeated friction or abrasion with harder materials. Rubber hardly ever slides on other rubber like materials but on tracks grossly dissimilar from it in surface texture, chemical constitution and elastic behavior, however, when the smooth surface is abraded, periodic parallel ridged patterns, looking like a wind-wrought pattern on sand, are formed on the rubber surface. These typical patterns are held through all processes of rubber abrasion, on the surface of tires, conveyor belts, printing rolls and shoes for example, which are thus regarded as the essential basis of rubber abrasion.

Abrasion process involves removal of small particles (1-5 μm) leaving behind pits in the surface and then followed by removal of large particles (> 5 μm) (El-Tayeb & Nasir, 2007). Detachment of small particles plays an important role in initiating the abrasion (Muhr & Roberts, 1992) and this is related to either a structural unit or localized stresses in the rubber. Since abrasion is clearly a manifestation of mechanical failure, Shallamach (Schallamach, 1557/58, 1968) used tearing energy to describe the rubber wear mechanisms, and Ratner et al. (Ratner et al., 1967) has established an equation in which the wear loss is related to macroscopic mechanical properties such as tensile strength, elongation at break, hardness etc. Thomas (Thomas, 1958, 1974) proposed the problem of abrasion is presented using fracture mechanics which treats fatigue and tensile failure as crack growth processes from small flaws. Crack growth can be influenced by the presence of oxygen or ozone. The nature of the vulcanizing system affects strength: crosslinks probably rupture and reform under stress. Other suggested factors responsible for particle detachment are internal subsurface failure due to flaws in the rubber such as dirt or voids (Gent, 1989) or interfacial adhesion at high speed rolling (Roberts, 1988). According to Pandey et al. (Pandey et al., 2003), there is no distinction between wear and abrasion, although other researcher (Schallamach, 1957/58) defined abrasion as that produced by laboratory machine on a rubber piece and wear as something that happens to tires or other rubber products. Thus, for rubber, "abrasion" covers all mechanisms, whereas the word "abrasion" for other materials refers in particular to scoring by hard, sharp particles. In the absence of transient effects such as clogging of the abrasive or evolution of an abrasion pattern it is found that the quantity of rubber abraded is proportional to the distance of sliding between rubber and counterface. However, wear of tires and abrasion on certain laboratory abrasion machines (e.g. the Akron abrader) brings into play gross properties of tire or test piece which affect the rate of wear just as much as does the abrasion resistance of the compound. However, Muhr and Roberts (Muhr & Roberts, 1992) proposed that abrasion reserves for processes where the amount of sliding is controlled, and that wear applies to the many practical situations for which the amount of sliding is as significant a variable as the attrition per unit sliding distance.

Schallamach (Schallamach, 1957/58, 1968) defines abrasion of rubber as a purely mechanical failure produced frictionally by the asperities of the track and this process creates periodic structures often called "abrasion pattern", a series of parallel ridges perpendicular to the sliding direction created on the surface of rubber during abrasive wear. He proposed the mechanism of rubber abrasion from a fracture mechanics point of view, relating the rate of wear to the crack growth resistance of the rubber. Although the concept of crack growth plays a very important role in abrasion, particularly in the growth of a single ridge, when we consider that the essential subject of rubber abrasion is in the formation of the periodic surface pattern consisting of very many cracks, moreover abrasive wear is a consequence of friction, in other words, it is impossible for any abrasive transfer of material to occur without friction phenomena. As is well known, an abrasion pattern is formed at the initial stage of abrasion and grows in ridge spacing and ridge height, whose geometric feature remains constant in appearance once it has grown up to the critical size. The abrasion pattern moves very slowly along in the sliding direction in a manner that the crack at the root of the pattern wedge is deepened somewhat and the protruding flap is tom off.

When rubber is slided over another abrasive surface, contact of abrasive grits (asperities) occurs. Upon application of a normal load, the extremely low tensile modulus of rubber ensures extensive deformation to establish a conformal contact with the counter part. This causes the real area of contact to become comparable to the apparent area of contact. Owing to the curved and entangled structure of chains of molecules in rubber, they can undergo considerable lateral deformation without fracture by stretching and twisting of chains. Gent (Gent, 1989) proposed a hypothetical mechanism for creating subsurface cracks during frictional sliding as part of the process of abrasive wear of rubber. It consists of the unbounded elastic expansion of microscopic precursor voids until they burst open as cracks, under the action of internal pressure or of a triaxial tension in the surrounding rubber. This conjecture accounts for enhanced resistance to abrasion for compounds reinforced with carbon black, in terms of increased stiffness without much loss of extensibility, and for the lack of correlation of abrasion resistance with other measures of strength. It should be noted that it is specific to soft, extensible materials, and thus it also accounts for marked differences in the nature of the wear process in rubbery materials compared to plastics and metals. Only rubbery materials appear to abrade away by a linking up of microcracks at right angles to the sliding direction to produce characteristic wear ridges known as the Schallamach abrasion pattern. Three mechanisms of generating a sufficiently large inflation pressure or triaxial tension are discussed. The most probable one seems to be thermal decomposition of rubber, generating volatile decomposition products a microscale blowout process. This would be aggravated by a simultaneous softening of the rubber on heating. Although strictly conjectural, it would be helpful to know whether the character of wear changes when abrasion is carried out under a large superimposed hydrostatic pressure. Sliding contact generates a shear stress along the surface of rubber and extensive shear stress along the stress axis is found to occur owing to the nature of elastomer (Chandrasekaran & Batchelor, 1997). When the shear stress exceeds the cohesive strength of the chain, fracture occurs by propagation of crack along the root of the contact area. The sudden release of deformation energy in the form of fracture in the contact area surface results in recovery of rubber to original coiled and entangled state. The visco-elastic nature of rubber limits the rate of recovery which results in shear wave propagation along the surface of rubber during sliding. Fukahori and Yamazaki (Fukahori & Yamazaki, 1994) proposed a new concept to understand the mechanism of formation of the periodic patterns characteristic in rubber abrasion. They showed that the driving force to generate the periodic surface patterns, and thus rubber abrasion consist of two kinds of periodic motion, stick-slip oscillation and the microvibration generated during frictional sliding of rubber. The stick-slip oscillation is the driving force to propagate cracks, then abrasion patterns and the microvibration with the natural frequency of the rubber induced in the slip phase of the stick-slip oscillation is another driving force for the initiation of the cracks. Although initial cracks originate in the slip region of the rubber surface, the propagation of the cracks is strongly excited in the stick region. Accordingly, the initial size of the abrasion pattern, pattern spacing, equals the distance determined by the natural period of the rubber and the sliding velocity while the constant pattern spacing after the critical number of frictional slidings agrees with the distance given by the period of the stick-slip oscillation and the sliding velocity. Consequently, during rubber abrasion, two driving forces produce bimodal size distribution of abraded particles, small particles of the order of ten micrometres by microvibrations and large ones of the order of a few hundred micrometres by the stick-slip motions. Sliding of rubber with high frictional forces does not necessarily entail abrasion (as it does for metal-metal contacts). Rather, abrasion of rubber results from mechanical failure due to excessively high local frictional stresses which are most likely to occur on rough tracks.

In spite of its practical importance, abrasion is perhaps the least understood phenomena amongst the various types of failures of rubber, as it is difficult to predict the abrasion behaviour from other rubber properties (Pandey et al., 2003). It is influenced by the hysteresis properties of the vulcanizates, magnitude of the frictional force and the resistance of rubber to rupture. The abrasion is a combination of mechanical, mechanochemical and thermochemical processes. The formation of an abrasion pattern depends on several factors such as crack growth process (Uchiyama and Ishino, 1992), mechanical properties of the rubber, and on the chemical, ageing and thermal conductivity properties of the composite. It depends on the modulus of elastomer, the abrasion pattern may be characterized by either ridges at the right angles to the rubbing direction if the modulus of elastomer is sufficiently low or score lines parallel to the rubbing direction if the modulus of elastomer is sufficiently high (Evstratov et al., 1967). There are a number of ways to make a material resistant to abrasion. One option is to utilize a special coating which creates a hardened layer over the material and resists friction. Some materials are also naturally extremely hard, and are ideal for settings in which abrasion resistance will be desirable. Other materials can be specifically formulated for increased hardness, as seen in plastics facilities which manufacture abrasion resistant plastics with the use of chemicals which harden and strengthen the plastic. The properties of a vulcanized rubber can be significantly influenced by details of the compounding. Practical materials will have, in addition to the base polymer, fillers, anti-degradants, crosslinking agents, accelerators etc. All of these can have an influence on the physical and chemical stability of the finished material.

2. Rubber use

2.1 Compounding

The rubber industry began when Charles Goodyear developed the first useful rubber compound: natural rubber plus sulfur. The concept of mixing materials into rubber to improve performance is still primary importance today. Without compounding, few rubbers would be of any commercial value. Any given rubber application will have a long list of necessary criteria in addition to cost, encompassing appearance, processing, mechanical, electrical, chemical, and thermal properties. Developing such compounds requires a broad knowledge of material science and chemistry combined with experience. The use of designed experiments can greatly facilitate selecting the optimum compound formulation.

The major components in a compound are rubber, vulcanizing agents, fillers, plasticizers, and antidegradants.

2.2 Rubber

Rubber is polymeric material endowed with the properties of flexibility and extensibility: with the application of force, the molecules straighten out in the direction in which they are being pulled; on release from being extended, they spontaneously recover their normal, random arrangements. Rubbers include natural rubber, a naturally occurring substance obtained from the exudations of certain tropical plants, and synthetic rubbers, artificially derived from petrochemical products. Among the most important synthetics are styrene-butadiene, polybutadiene, and polyisoprene (commonly classified as the "general purpose"), as well as ethylene-propylene rubber (often referred to as "specialty rubber"). The prices of these synthetics have been historically in the range of natural rubber prices and their markets have, although to varying degrees, overlapped those of natural rubber. Synthetic rubbers are materials with distinctive chemical structures, whereas the emphasis with natural rubber lies on different types and grades within one single broad category.

2.2.1 Natural rubber

Natural rubber (polymer designation cis-1-4 polyisoprene, empirical formula $(C_5H_8)_n$ obtained from the sap (latex) of several rubber-yielding plants (e.g. Hevea Brasiliensis and

Parthenia argentatum) by coagulation with chemicals, drying, electrical coagulation, and other processes is the prototype of all elastomers. Latex extracted in the form of latex from the bark of the Hevea tree is a polydispersed colloidal system of rubber particles in an aqueous phase. With Hevea latex, the dry rubber content varies between approximately 28-40%, although it may rise to 45-50% after a long period of non-tapping (notice that, for statistical purposes, figures for natural rubber may include the dry rubber content of latex). The rubber produced from latex contains, besides the hydrocarbon, relatively small quantities of protein, carbohydrates, resin-like substances, mineral salts, and fatty acids.

The natural rubber polymer is nearly 100% cis-1,4 polyisoprene with Mw ranging from 1 to 2.5 x 10^6. Due to its high structural regularity, natural rubber tends to crystallize spontaneously at low temperatures or when it is stretched. Low temperature crystallization causes stiffening, but is easily reversed by warming. The strain-induced crystallization gives natural rubber products outstanding resilience, flexibility, and tear and tensile strength, as well as low heat build-up and abrasion. However, a drawback is natural rubber moderate environmental resistance to factors such as oxidation and ozone; so too for its scarce resistance to chemicals, including gasoline, kerosene, hydraulic fluids, degreasers, synthetic lubricants, and solvents. In addition, latex contains proteins that can cause severe allergic response in a small percentage of the population and among medical professionals following extensive exposure. The largest use of natural rubber is in the tires. Over 70 percent of its manufacture consumption is in this area. The next largest use is as latex in dipped goods, adhesives, rubber thread, and foam. These uses account for approximately another 10 percent. The of applications remainder is used in a variety such as convey or belts, hoses, gaskets, footwear, and antivibration devices such as engine mounts.

2.2.2 Polyisoprene

Synthetic polyisoprene is designed to be similar to natural rubber in structure and properties. Although it still demonstrates lower green strength, slower cure rates, lower hot tear, and lower aged properties than its natural counterpart, synthetic polyisoprene exceeds the natural types in consistency of product, cure rate, processing, and purity. In addition, it is superior in mixing, extrusion, molding, and calendering processes. The lithium based polymers were found to produce up to 94 percent cis, which still was not high enough to provide the properties of natural rubber. Polymers made with the coordination catalysts have cis contents of up to 98 percent, providing products that can more closely serve as replacements for natural rubber than the lithium-based polymers. In comparison with natural rubber, they offer the advantage of a more highly pure rubber (no non-rubber material) and excellent uniformity. A high trans-1,4 structure was produced by Polysar, and is now being produced by Kuraray. A Li-IR with increased 3,4 structure can be prepared by adding polar modifiers to the alkyl lithium catalyst system. However, since the higher cis-1,4 configuration most closely mirrors the properties of natural rubber and is the most important commercially. Currently synthetic polyisoprene is being used in a wide variety of industries in applications requiring low water swell, high gum tensile strength, good resilience, high hot tensile, and good tack. Gum compounds based on synthetic polyisoprene are being used in rubber bands, cut thread, baby bottle nipples, and extruded hose. Carbon black loaded compounds find use in tires, motor mounts, pipe gaskets, shock absorber bushings and many other molded and mechanical goods. Mineral filled systems find applications in footwear, sponge, and sporting goods. In addition, recent concerns about allergic reactions to proteins present in natural rubber have prompted increased usage of the more pure synthetic polyisoprene in some applications.

2.2.3 Styrene-butadiene rubber

The largest-volume synthetic rubber consumed is styrene-butadiene rubber (SBR) produced by both emulsion (E-SBR) and solution (S-SBR). In 2003, SBR solid rubber accounted for 4% percent of all synthetic rubber. If SBR latex and carboxylated SBR latex are included, its share increases to 55 percent. The major application of solid SBR is in the automotive and tire industry, accounting for approximately 70 percent of the use. Therefore, SBR has been tightly tied to the tire business.

Most of the E-SBR contains about 24% by weight of styrene and it is a random copolymer with butadiene. Some specific grades contain as much as 40-46% styrene, and are much stiffer. The polymerization is by free radical initiator and there is a finite probability of chain-transfer reaction, which generates long branches.

Emulsion polymerization is carried out either hot, at about 25-50°C, or cold, at about 5-25°C, depending upon the initiating system used. SBR made in emulsion usually contains about 24% styrene randomly dispersed with butadiene in the polymer chains. At high temperature polymerization, many long-braches and gels were formed. The rubber was stiff and difficult to mill, mix, or calender than natural rubber, deficient in building tack, and having relatively poor inherent physical properties. Processability and physical properties were found to be greatly improved by the addition of process oil and reinforcing pigments. Polymerization at a lower temperature became possible, giving less branches and gels. "Cold" SBR generally has a higher average molecular weight and narrower molecular weight distribution. It thereby offers better abrasion and wear resistance plus greater tensile and modulus than "hot" SBR. Since higher molecular weight can make cold SBR more difficult to process, it is commonly offered in oil-extended form. S-SBR comes in two distinctly different subgroups, one made by an anionic initiator and the other by free radical initiators. SBR made in solution contains about the same amount of styrene, but both random and block copolymers can be made. Solution SBR can be tailored in polymer structure and properties to a much greater degree than their emulsion counterparts. The random copolymers offer narrower molecular weight distribution, low chain branching, and lighter color than emulsion SBR. They are comparable in tensile, modulus, and elongation, but offer lower heat buildup, better flex, and higher resilience. Certain grades of solution SBR even address the polymer's characteristic lack of building tack, although it is still inferior to that of natural rubber. The processing of SBR compounds in general is similar to that of natural rubber in the procedures and additives used. SBR is typically compounded with better abrasion, crack initiation, and heat resistance than natural rubber. SBR extrusions are smoother and maintain their shape better than those of natural rubber. SBR was originally developed as a general purpose elastomer and it still retains this distinction. It is the largest volume and most widely used elastomer worldwide. Its single largest application is in passenger car tires particularly in tread compounds for superior traction and tread wear. Substantial quantities are also used in footwear, foamed products, wire and cable jacketing, belting, hoses, and mechanical goods.

2.2.4 Polybutadiene rubber

Polybutadiene rubber (BR) was originally made by emulsion polymerization, generally with poor results. It was difficult to process and did not extrude well. This rubber became commercially successful only after it was made by solution polymerization using stereospecific Ziegler-Natta catalysts. This provided a polymer with greater than 90% cis-1,4-polybutadiene configuration. This structure hardens at much lower temperatures (with T_g of -100°C) than natural rubber and most other commercial rubbers. This gives better low temperature flexibility and higher resilience at ambient temperatures than most rubbers. Greater resilience means less heat buildup under continuous dynamic deformation as well. This high-cis BR was also found to possess superior abrasion resistance and a great tolerance for high levels of extender oil and carbon black. High-cis BR was originally blended with natural rubber simply to improve the latter's processing properties, but it was found that the BR conferred many of its desirable properties to the blend. The same was found to be true in blends with SBR.

The 1,3-butadiene monomer can polymerize in three isomeric forms: by cis 1,4 addition, trans 1,4 addition, and 1,2 addition leaving a pendant vinyl group. By selection of catalyst and control of processing conditions, polybutadiene is now sold with various distributions of each isomer within the polymer chain, and with varying levels of chain linearity, molecular weight and molecular weight distribution. Each combination of chemical properties is designed to enhance one or more of BR's primary attributes.

The major use of polybutadiene (cis-1,4-BR) having a very low glass transition temperature in the region -75°C to -100°C is in tires, with over 70 percent of the polymer produced used by the tire industry, primarily in blends with SBR or natural rubber to improve hysteresis (resistance to heat buildup), abrasion resistance, and cut growth resistance of tire treads. The type of BR used depends on which properties are most important to the particular compound. High-cis and medium-cis BR have excellent abrasion resistance, low rolling resistance, but poor wet traction. High-vinyl BR offer good wet traction and low rolling resistance, but poor abrasion resistance. Medium-vinyl BRs balance reasonable wet traction with good abrasion resistance and low rolling resistance. Polybutadiene is also used for improved durability and abrasion and flex crack resistance in tire chaffer, sidewalls and carcasses, as well as in rubber blends for belting. High- and medium-cis BRs are also used in the manufacture of high impact polystyrene. Three to twelve percent BR is grafted onto the styrene chain as it polymerizes, conferring high impact strength to the resultant polymer.

Polybutadiene made by emulsion polymerization with a free radical initiator is used as the rubber component of an impact modifier in plastics, in particular high impact polystyrene (HIPS) and acrylonitrile-butadiene-styrene resin (ABS). In the HIPS application the rubber is dissolved in the styrene monomer, which is then polymerized via a free-radical mechanism. A complex series of phase changes occurs, resulting in small rubber particles containing even smaller polystyrene particles being incorporated into a polystyrene matrix. The rubber is added to increase impact strength. Because of the unique morphology that is formed, low levels of rubber (typically around 7%) provide rubbery particles having a volume fraction of 30-40 percent. This morphology leads to high impact at very low rubber levels, providing good stiffness and hardness.

There is also a fairly large market for high cis BR in solid core golf balls. In this application, the polymer is compounded with zinc acrylate and the mixture is cured with peroxide. This produces an ionically cross-linked compound that has outstanding resilience. The covers are also ionomers with superior cut resistance. In the last few years the golf ball market has been shifting away from the traditional wound ball to these new solid core balls that use polybutadiene.

2.2.5 Nitrile rubber

Acrylonitrile butadiene rubber (NBR) is made as an emulsion with a free radical initiator. Polymers are made with an acrylonitrile (AN) content of, for example, 28, 33 or 40 weight percent, depending upon the required oil resistance. It also has good elongation properties as well as adequate resilience, tensile and compression set. The major applications for this material are in areas requiring oil and solvent resistance. As the acrylonitrile content increases in the polymer chain, the properties change predictably. The glass transition temperature increases approximately 1.5°C for each percent increase in acrylonitrile. Properties such as hysteresis loss, resilience, and low-temperature flexibility will correspondingly change. The oil resistance increases with increased acrylonitrile content, as does the compatibility with polar plastics such as PVC. The major market for nitrile rubber is in the automotive area because of its solvent and oil resistance. Major end uses are for hoses, fuel lines, O-rings, gaskets, and seals. In blends with polyvinyl chloride (PVC) and acrylonitrile butadiene styrene (ABS), nitrile rubber acts as an impact modifier. Some nitrile rubber is sold in latex form for the production of grease-resistant tapes, gasketing material, and abrasive papers. Latex also is used to produce solvent resistant gloves.

Hydrogenated NBR (HNBR) is produced by first making an emulsion-polymerized NBR using standard technique. Almost all the butadiene units become saturated to produce an ethylene-butadiene-acrylonitrile terpolymer. These "post-polymerization" reactions are very expensive so HNBR usually command a premium price. HNBR is usually cured with peroxides, similar to ethylene-propylene rubber, because it has no unsaturation for a conventional sulfur cure system. HNBR has many uses in the oil-field, including down hole packers and blow-out preventers, because of its outstanding oil resistance and thermal stability. For the same reasons, it has also found uses in various automotive seals, O-rings, timing belts, and gaskets. Resistance to gasoline and aging make HNBR ideal for fuel-line hose, fuel-pump and fuel injection components, diaphragms, as well as emission-control systems. HNBR is the best selection to achieve to the highest abrasion and heat resistance. Service temperature of this rubber is up to 160°C and it's used in temperatures up to 200°C for short times and its abrasion resistance and wet traction is very good. But price of HNBR is high and it isn't an economy rubber for general applications.

2.2.6 Ethylene-propylene rubber

Ethylene-propylene rubber continues to be one of the most widely used and fastest growing synthetic rubber having both specialty and general purpose applications. Polymerization and catalyst technologies in use today provide the ability to design polymers to meet specific and demanding application and processing needs. Versatility in polymer design and performance has resulted in broad usage in automotive weather-stripping and seals, glass-run channel, radiator, garden and appliance hose, tubing, belts, electrical insulation, roofing membrane, rubber mechanical goods, plastic impact modification, thermoplastic vulcanizates and motor oil additive applications. Ethylene-propylene rubber are valuable for their excellent resistance to heat, oxidation, ozone and weather aging due to their stable, saturated polymer backbone structure. Properly pigmented black and non-black compounds are color stable. As non-polar rubber, they have good electrical resistivity, as well as resistance to polar solvents, such as water, acids, alkalies, phosphate esters and many ketones and alcohols. Amorphous or low crystalline grades have excellent low temperature flexibility with glass transition points of about minus 60°C. Ethylene-propylene rubber uses the same chemical building blocks or monomers as polyethylene (PE) and polypropylene (PP) thermoplastic polymers. These ethylene (C2) and propylene (C3) monomers are combined in a random manner to produce rubbery and stable polymers. There are two general types of polymers based on ethylene and propylene: ethylene-propylene rubber (EPM) and ethylene-propylene terpolymer (EPDM). EPM accounts for approximately 20 percent of the polyolefin rubber produced. Comprising a totally saturated polymer, these materials require free-radical sources to cross-link. EPDM was developed to overcome this cure limitation. For EPDM a small amount (less than 15%) of a nonconjugated diene is terpolymerized into the polymer. One of the olefinic groups is incorporated into the chain, leaving its other unsaturated site free for vulcanization or polymer modification chemistry. This ensures that the polymer backbone remains saturated, with corresponding stability, while still providing the reactive side group necessary for conventional cure systems. The nonconjugated dienes used commercially are ethylidene norbornene, 1,4 hexadiene, and dicyclopentadiene. Each diene incorporates with a different tendency for introducing long chain branching (LCB) or polymer side chains that influence processing and rates of vulcanization by sulfur or peroxide cures.

2.3 Vulcanizing agents

Vulcanization is a chemical process for converting rubber or related polymers into more durable materials via the addition of sulfur or other equivalent vulcanizing agent. The function of vulcanizing agent is to modify the polymer by forming crosslinks (bridges) between individual polymer chains; the most common ones are the sulfur type for unsaturated rubber and peroxides for saturated polymers. Uncured natural rubber is sticky, deforms easily when warm, and is brittle when cold. In this state, it is a poor material when a high level of elasticity is required. Vulcanized material is less sticky and has superior mechanical properties. A vast array of products is made with vulcanized rubber including tires, shoe soles, hoses, and hockey pucks. The main polymers subjected to vulcanization are polyisoprene (natural rubber) and styrene-butadiene rubber, which are used for most passenger tires. Chemicals called accelerators may be added to control the cure rate in the sulfur system; these materials generally are complex organic chemicals containing sulfur and nitrogen atoms. Stearic acid and zinc oxide usually are added to activate these accelerators. Metal oxides are used to cure halogenated polymers such as polychloroprene or chlorosulfonated polyethylene.

2.4 Fillers

Natural and synthetic rubbers, also called elastomers are rarely applied in their pure form. They are too weak to fulfill practical requirements because of lack of hardness, strength properties and wear resistance. A rubber compound contains, on average, less than 5 kg of chemical additives per 100 kg of rubber, while filler loading is typically 10-15 times higher. Of the ingredients used to modify the properties of rubber products, the filler often plays a significant role. Most of the rubber fillers used today offer some functional benefit that contributes to the processability or utility of the rubber product. Styrene butadiene rubber, for example, has virtually no commercial use as an unfilled compound. Fillers are used in order to improve the properties of rubber compounds. The characteristics which determine the properties a filler will impart to a rubber compound are particle size, particle surface area, particle surface activity and particle shape. Surface activity relates to the compatibility of the filler with a specific rubber and the ability of the rubber to adhere to the filler. Rubber articles derive many of their mechanical properties from the admixture of these reinforcing (active) fillers at quantities of 30% up to as much as 300% relative to the rubber part. The introduction of carbon black as a reinforcing agent in 1904, lead to strongly increased tread wears resistance. Carbon black is in use as the most versatile reinforcing filler for rubber, complemented by silicas. In tire manufacturing silicas are more and more used nowadays, mainly to decrease the rolling resistance. The increased attitude of protecting the environment gives rise to a demand for tires combining a long service life with driving safety and low fuel consumption, achieved by this lower rolling resistance. However, the change from carbon black to silica is not at all obvious because of technical problems involved. In particular, the mixing of rubber with pure silicas is difficult, because of the polarity-difference between silica and rubber. Therefore, coupling agents are applied in order to bridge this polarity difference. Sometimes fillers are added to reduce cost, increase hardness, and color the compound. Generally they do not provide the dramatic improvement in properties seen with reinforcing agents, but they may have some reinforcing capability. Carbon black and silica are the most common reinforcing agents. These materials improve properties such as tensile strength and tear strength; also, they increase hardness, stiffness, and density and reduce cost. Almost all rubbers require reinforcement to obtain acceptable use properties. The size of the particles, how they may be interconnected (structure), and the chemical activity of the surface are all critical properties for reinforcing agents. In tire applications, new polymers are currently being developed which contain functional groups that directly interact with carbon black and silica, improving many properties. Typical fillers are clays, calcium carbonate, and titanium dioxide.

2.5 Plasticizers

These materials are added to reduce the hardness of the compound and can reduce the viscosity of the uncured compound to facilitate processes such as mixing and extruding. The most common materials are petroleum-based oils, esters, and fatty acids. Critical properties of these materials are their compatibility with the rubber and their viscosity. Failure to obtain sufficient compatibility will cause the plasticizer to diffuse out of the compound. The oils are classified as aromatic, naphthenic, or paraffinic according to their components. Aromatic oils will be more compatible with styrene-butadiene rubber than paraffinic oils, whereas the inverse will be true for butyl rubber. The aromatic oils are dark colored and thus cannot be used where color is critical, as in the white sidewall of a tire. The naphthenic and paraffinic oils can be colorless and are referred to as nonstaining.

2.6 Antidegradents

An antidegradant, this group of chemicals is an ingredient in rubber compounds to deter the aging of rubber products. The most important are the antioxidants, which trap free radicals and prevent chain scission and crosslinking. Antiozonants are added to prevent ozone attack on the rubber, which can lead to the formation and growth of cracks. Antiozonants function by diffusion of the material to the surface of the rubber, thereby providing a protective film. Certain antioxidants have this characteristic, and waxes also are used for this purpose.

2.7 Processing

A wide range of processes are used to convert a bale of rubber into a rubber product such as a tire. The first process generally will be compounding. Typical compounding ingredients were discussed previously. In many compounds more than one rubber may be needed to obtain the performance required. Uncured rubber can be considered as a very high viscosity liquid; it really is a viscoelastic material possessing both liquid and elastic properties. Mixing materials into rubber requires high shear, and the simplest method is a double roll mill in which the rubber is shear-mixed along with the other compounding ingredients in the bite of the mill. Large scale mixing is most commonly done with a high-shear internal mixer called a Banbury. This mixing is a batch process, although continuous internal mixers also are used. The compounded rubber stock will be further processed for use. The process could be injection or transfer molding into a hot mold where it is cured. Tire curing bladders are made in this fashion. Extrusion of the rubber stock is used to make hose or tire treads and sidewalls. Another common process is calendaring, in which a fabric is passed through rolls where rubber is squeezed into the fabric to make fabric-reinforced rubber sheets for roofing membranes or body plies for tires. The actual construction of the final product can be quite complex. For example, a tire contains many different rubber components some of which are cord or fabric reinforced. All of the components must be assembled with high precision so that the final cured product can operate smoothly at high speeds and last over 50,000 miles.

3. Abrasion test

An abrasion test is a test used to measure the resistance of a material to wear stemming from sliding contact such as rubbing, grinding, or scraping against another material. Abrasion may be measured in a variety of ways, depending on the resistance test used and the information that is desired from the test (Dick, 2001). For example, where the amount of material lost is a concern regardless of whether the material fails, abrasion may be measured in terms of the percentage of material lost, either by mass or by volume, between the start and end of the test. Another measure sometimes used is the number of abrasion cycles a material withstands before failure. This would be more appropriate if information on how long the material or product will survive before outright failure is of primary interest. Abrasion tests try to accelerate the process by applying more cutting-like conditions; however, this approach may not simulate actual wear. It is also important to try to match the severity of the abrasion test to the severity of the product wear conditions. For example, the severity of test conditions imparted by most abraders is usually greater than what the highway pavement may impart to a tire tread compound during normal driving. Wear resistance is an important rubber compound property related to the useful product life for tires, belts, shoe soles, rubber rolls, and sandblasting hose, among other products. A wide variety of different abrasion testers have been developed over the years in an attempt to correlate to these product wear properties. Several factors are typically considered in developing or selecting an appropriate abrasion test for the application at hand. The shape of the contact area is taken into consideration, as is the composition of the two surfaces making contact with one another. Speed of sliding contact between the two surfaces, the force with which they act on one another, and the duration of contact between them may also be considered. In addition to the materials themselves, the environment in which they are making contact also plays a role in selecting an appropriate abrasion test.

The abrasion resistance is expressed as volume loss in cubic millimetres or abrasion resistance index in percent. For volume loss, a smaller number indicates better abrasion resistance, while for the abrasion resistance index; a smaller number denotes poorer abrasion resistance. Tested compounds are usually compared on a "volume loss" basis which is calculated from the weight loss and density of the compound. Abrasion test results are known to be variable; therefore, it is important to control and standardize the abradant used in the test. Also, it is a good idea to relate test results to a standard reference vulcanizates.

ASTM D394, the Dupont Abrasion Test Method, consists of a pair of rubber test pieces pressed against a disk of a specified abrasive paper which rotates whilst a pair of moulded test pieces is continuously pressed against it either with a constant force or with a force adjusted to give a constant torque on the arm holding the test pieces. Care should be taken with soft rubber compounds because "smearing" can occur, affecting test results.

ASTM D1630 describes the rotary-platform, double-head abrader is commonly referred to as the NBS Abrader used on rubber compounds for shoe soles and heels. The NBS abrader uses rotating drums with a specified abrasive paper around them onto which the test pieces are pressed by means of levers and weights, a specified standard reference compound to be used for the calculation of an abrasive index.

ASTM D2228 describes the Pico Abrader. This unique test works on the principle of abrading the rubber surface by rotating a rubber specimen against a pair of tungsten carbide knives. A special dusting power is fed to the test piece surface, which doubtless helps to avoid stickiness. This method specifies five standard rubbers and the result also expressed as an abrasion index. Force on the test piece and speed of rotation can be varied and, presumably, different abradant geometries could be used, although the distinctive feature of the Pico is the use of blunt metal knives in the presence of a powder.

ASTM D3389 refers to the Taber Abrader using a pair of abrasive wheels, a method not originally from the rubber industry. This very general method uses two abrasive wheels against the rubber test piece (disk) attached to a rotating platform. Although the degree of slip cannot be varied; however, the force on the test piece and the nature of the abradant are very readily varied and tests can be carried out in the presence of liquid or powder lubricants. When using the usual type of abrasive wheel, a refacing procedure is carried out before each material tested.

ISO 4649 refers to the DIN Abrader, based on the German Standard. The rubber test piece with a holder is traversed a rotating cylinder covered with a sheet of the abradant paper. By allowing the sample holder to move the test piece across the drum as it rotates, there is less chance of rubber buildup on the abradant paper. This method, used extensively in Europe, is very convenient and rapid and well suited to quality control the uniformity of a specific material. The achieved test results provide important parameters in respect to the wear of rubbers in practical use. The details of procedure and expression of results are something of a compromise, being a compilation of the German approach and the British approach. Two procedures are specified, using a rotating or non-rotating test piece respectively. In principle, the abrasion should be more uniform if the test piece is rotated during test. The standard abradant is specified in terms of weight loss of a standard rubber using a non-rotating test piece and has to be run in against a steel test piece before use. Results can either be expressed as a relative volume loss with the abradant normalized relative to a standard rubber or as an abrasion index relative to a standard rubber.

British Standard BS903: Part A9 still describes the Akron Abrader. The rubber test piece is a moulded wheel which is positioned against an abrasive cylinder under constant speeds and held against the abrasive wheel by a constant force. The Akron Abrader has the advantage of allowing variation in the degree of slip in the test by varying the angle of the test piece.

4. Effect of compounding ingredients on abrasion resistance

4.1 Rubber

In rubbery materials, when the smooth surface is abraded, periodic parallel ridged patterns are formed on the rubber surface. These typical patterns are held through all processes of rubber abrasion, on the surface of tires, conveyor belts, printing rolls and shoes for example, which are thus regarded as the essential basis of rubber abrasion. In the absence of any serious chemical decomposition the abrasion process initially results in the removal of small rubber particles just a few microns in size, leaving pits behind in the surface. With continued rubbing, larger pieces of rubber are removed. Although most weight loss is attributable to the larger pieces, it is thought that the detachment of the smaller particles initiates the abrasion process. The small particles have a characteristic size of 1-5 μm, but whether this relates to a structural unit in the rubber compound (Muhr & Roberts, 1992). Other suggestions are that mechanical rupture to produce the particles relates to flaws in the rubber, including dirt, or voids that cavitate leading to internal subsurface failure (Gent, 1989). A rolling experiment suggested that particle detachment might be linked to interfacial adhesion (Roberts, 1988). Schallamach (Schallamach, 1557/58, 1968) reported that rubber often develops a pattern of ridges perpendicular to the direction of abrasion. In the simplest case abrasion is produced by a line contact pulling a tongue of rubber from the ridge producing crack growth at the base of the tongue. Provided the surface configuration is in a steady state, the quantity of rubber abraded can be related quantitatively to the frictional force and the crack growth characteristic of the rubber. The abrasion of rubber results from mechanical failure due to excessively high local frictional stresses which are most likely to occur on rough tracks. Theories of abrasion thus require details of the local stresses, which together with the strength properties of the rubber may enable the rate of abrasion to be predicted. Gent and Pulford (Gent & Pulford, 1983) reported the reversal in the relative rates of wear of unfilled polybutadiene rubber comparing to those of unfilled natural rubber and styrene butadiene rubbers as frictional force increased. Fukahori and Yamazaki (Fukahori & Yamazaki, 1994) investigated the mechanism of the formation of the periodic ridges in rubber abrasion by designing the razor blade abrader. They reported that the driving force to generate the periodic surface patterns, and thus rubber abrasion consists of two kinds of periodic motions, stick-slip oscillation and the microvibration generated during frictional sliding of rubber. The stick-slip oscillation is the driving force to propagate cracks, then abrasion patterns and the microvibration with the natural frequency of the rubber induced in the slip phase of the stick-slip oscillation is another driving force for the initiation of the cracks. Although initial cracks originate in the slip region of the rubber surface, the propagation of the cracks is strongly excited in the stick region. Champ et al. (Champ et al., 1974) proposed the mechanism of rubber abrasion from a fracture mechanics point of view, relating the rate of wear to the crack growth resistance of the rubber. Although the concept of crack growth plays a very important role in abrasion, particularly in the growth of a single ridge, when consider that the essential subject of rubber abrasion. Liang et al. (Liang et al., 2010) investigated the blade abrasion of four different rubber materials, unfilled natural rubber, unfilled styrene butadiene rubber, unfilled polybutadiene rubber and carbon black filled styrene-butadiene rubber. Each is abraded until the steady state abrasion pattern is developed on the surface of moulded rubber wheels. The steady state conditions are measured using the weight loss per revolution of the wheel. The abraded surface is cut to examine the typical asperity profile. Each profile is modeled using finite element analysis to calculate the stored energy release rate for each combination of material and test condition. The stored energy release rate when combined with an independent measure of the rate of crack growth measured using a fatigue crack growth test gives a reasonable prediction of the abrasion rate. They has shown that the low strength of the BR material results in much smaller asperities being formed under steady state abrasion which results in a much slower abrasion rate. Conversely the strongest material NR has the longest tongue on the asperity and this in turn generates much larger values for the tearing energy at the tip of the asperity and this contributes to its poor abrasion resistance. Hong et al. (Hong et al., 2007) observed that BR compounds caused much slower wear than NR and SBR compounds. Arayapranee and Rempel (Arayapranee & Rempel, 2009) studied the cure characteristics, mechanical properties before and after heat ageing, and abrasion and ozone resistances of hydrogenated natural rubber (HNR), providing an ethylene–propylene alternating copolymer, vulcanizate and compared with those of natural rubber (NR), ethylene propylene diene terpolymer (EPDM) and 50:50 NR/EPDM vulcanizates. They reported that the highest abrasion resistance of the NR vulcanizate could be attributed to high unsaturated structure, as evident from its highest tensile strength compared to other vulcanizates. The abrasion loss of 48% HNR is higher than that of the NR vulcanizate, due to a reduction in the number of the double bonds. This suggests that abrasion resistance is heavily dependent on the unsaturation content in the backbone chain.

4.2 Fillers

Fillers increase the stiffness of rubber in various degrees depending on quantity and quality of the fillers. The properties of rubber compounds are affected not only by the filler content but also by its structure and particle size. Despite outstanding resilience and high tensile strength, natural rubber possesses poor abrasion resistance. Thus, blending with high abrasion resistance rubbers and/or reinforcing by inorganic fillers are generally used to improve the abrasion resistance of NR and other rubbers. Carbon black and silica are two common fillers used to reinforce rubbers. However, high loadings of these fillers are required to obtain desirable properties. Incorporation of reinforcing fillers such as carbon black improves stiffness and strength of rubber (Tabsan et al., 2010). Hence, the abrasion resistance is improved by suppressing tearing of the rubber under the sliding contact (Gent & Pulford, 1983). Arayapranee at el. studied the effect of filler type and loading on the abrasion loss (Arayapranee at el., 2005). They found that the incorporation of silica and carbon black reduces the abrasion loss of the natural rubber materials notably, whereas rice husk ash shows no effect with filler loading. Reinforcing fillers, silica and carbon black, interact preferentially with the natural rubber phase, as shown by the higher reduction of abrasion loss in the compounds. This improvement is probably due to the greater surface area and better filler-rubber interfacial adhesion resulting in an improved abrasion resistance. Fine particles actually reflect their greater interface between the filler and the rubber matrix and, hence, provide a better abrasion resistance than the coarse ones. Similar results were also reported by (Sae-oui et al., 2002).

Filled compounds are found to be less sensitive to the frictional force, whether wear took place by tearing or by smearing (Gent & Pulford, 1983). Carbon black is an additive with a decisive effect on the abrasion resistance. Hong et al. (Hong et al., 2007) investigated the effects of the particle size and structure of various carbon blacks on friction and abrasion behavior of filled natural rubber, styrene–butadiene rubber and polybutadiene rubber using a modified blade abrader. The effect of particle size and structure on abrasion resistance should be considered for the optimum design of desired wear properties. The worn surfaces of the rubber compounds filled with carbon black having smaller particle size and a more developed structure showed narrower spaced ridges and better abrasion resistance. It means that smaller particle size and better structure development of carbon black resulted in improved abrasion resistance. Yang et al. (Yang et al., 1991) reported that the abrasive wear of rubbers is strongly affected by the filler particles dispersed in the rubber matrix. The fillers are incorporated usually for the purposes of mechanical reinforcement and improving the conductivity of the neat resins. It is found that rigid filler particles normally increase the abrasive wear loss of the filled silicone rubbers. The wear rates of the filled silicone rubbers increase slowly with filler concentration until a critical volume fraction is reached, at which point they increase very rapidly with increasing filler. The critical filler fraction should carry important information, as it apparently divides two wear regimes dominated by different mechanisms. The first regime, where the filler concentration is low, is dominated by the properties of the neat resin. The increase of wear rate due to the filler is gradual here. In the cases of effective filler reinforcement, a reduction of wear rate can occur. The second regime is dominated by the filler's detrimental effects where the wear rate increases very rapidly with filler concentration. The stress concentration introduced by the rigid particles effectively creates a 'damage zone' surrounding the particles, a location where micro-cavitation and debonding takes place. Cavitation appears to dominate in the composites of very small filler particles, while debonding dominates when larger particles are involved. In view of the importance of carbon blacks on tread wear, it is surprising that relatively little understanding of the phenomenon has been set out in print. Although for synthetic rubbers such as BR and SBR it may seem unnecessary to look further than the dramatic enhancement of strength properties imparted by the use of particular grades of black, for NR such enhancement is modest and additional mechanisms for the effect of blacks on tread wear should be sought. In any case there is a consensus that high surface area, high surface activity and high structure promote tread wear resistance. Even so, the evidence that carbon black does not necessarily enhance the abrasion resistance of rubber under conditions of equal sliding suggests that the effect of carbon black on tread wear may in part be simply associated with stiffening, and hence reduced sliding, without weakening the compound as a high crosslink density would do. Arayapranee and Rempel (Arayapranee & Rempel, 2008) studied the effects of incorporation of three different fillers, i.e. rice husk ash (RHA), silica and calcium carbonate ($CaCO_3$), over a loading range of 0-60 phr on the abrasion loss of 75:25 natural rubber (NR)/ethylene propylene diene monomer (EPDM) blends. The incorporation of silica reduced the abrasion loss of the 75:25 NR/EPDM blends notably, whereas $CaCO_3$ showed a different trend in abrasion loss tending to increase it with an increase in $CaCO_3$ loading. However, RHA showed less of an effect with filler loading. At a similar filler loading, silica filled 75:25 NR/EPDM blends had the lowest abrasion loss followed by RHA and $CaCO_3$ filled 75:25 NR/EPDM blends.

4.3 Lubricants

Lubricants is widely used in the compounding of diene rubbers to improve the processability of the compounds and to impart the desired physicomechanical properties of rubber compounds and vulcanizates. The presence of a liquid can prevent moving surfaces from coming into inimate contact if viscous flow from the contact region is sufficiently sluggish. Lubricants, such as non-swelling fluids or dust, greatly reduce friction on smooth surface but the effect is smaller on rough surface. Changes in friction properties of rubber are possible by adding substantial amounts of standard lubricants, but this reduces strength, especially at high temperatures. Contrary, improvement in friction properties of rubbers based on blends NR and BR could be reached by introducing 0.5 wt% of K95 experimental lubricant (Jurkowska et al., 2006). Lubricant K95 added in a quantity of 0.5 wt% reduced the viscosity of rubber compound; it also improved compound flow in the mold. Mechanical properties of cured rubber not decrease while resistance to abrasion and fatigue increased. The influence of Lubricant K95 on reducing of the internal friction of rubbers is found.

Evstratov et al. (Evstratov et al., 1967) found that abrasion on a ribbed metal surface increases abruptly, by an order of magnitude or so, when the friction coefficient (μ) exceeds about 1.4. Abrasion patterns were observed for μ above the critical value, but not for lower values. It did not matter whether μ was an unlubricated value for the compound or was determined by the presence of a lubricant. The renowned abrasion resistance of cis-BR compounds may relate to this observation; such compounds have low dry friction and form only very fine abrasion patterns. In spite of their low strength, their abrasion resistance can be excellent. When a lubricant is applied, a much finer pattern develops and the rate of abrasion is much lower.

4.4 Antioxidants

Gent and Pulford (Gent & Pulford, 1983) determined rates of wear have been determined for several rubber materials, using a razor-blade abrading apparatus at different levels of frictional power input, corresponding to different severities of wear, at both ambient temperature and at 100°C, and both in air and in an inert atmosphere. It is concluded that wear occurs as a result of two processes: local mechanical rupture (tearing) and general decomposition of the molecular network to a low-molecular-weight material (smearing). The decomposition process could, in principle, be ascribed to several mechanisms: thermal decomposition due to local heating during sliding; oxidative deterioration, possibly accelerated by local heating; and mechanical rupture of macromolecules to form reactive radical species. The most plausible mechanism of smearing appears to be oxidative consummation of scissions produced by mechanical stress, in much the same way as occurs during cold mastication of natural rubber. They provided rather convincing evidence of mechanochemical degradation of certain rubbers during abrasion by a razor blade. The degradation of carbon black-filled natural rubber (NR), styrene-butadiene rubber (SBR), and ethylene-propylene rubber (EPM) to a sticky material during blade abrasion occurred only in the presence of oxygen or thiophenol, but not in a nitrogen atmosphere (just as for cold mastication). Polybutadiene rubber (BR) produced only dry debris during abrasion, consist with the expectation that any free radicals of BR produced by main chain rupture would react with the polymer itself, leading to an increase in cross-linking rather than degradation. Carbon black-filled natural rubber, styrene-butadiene rubber, and ethylene-propylene rubber were particularly susceptible to decomposition and smearing, but for natural rubber and SBR the decomposition process was not observed in an inert atmosphere. It is attributed to molecular rupture under frictional forces followed by stabilization of the newly formed polymeric radicals by reaction with oxygen, if present, or with other polymer molecules, or with other macroradicals. Polybutadiene rubber produced only dry debris during abrasion. Radicals of BR produced by main chain rupture would react with the polymer itself, leading to an increase in cross-linking rather than degradation. Rates of wear have been found to increase with the applied frictional force raised to a power n. The value of n was between 2.5 and 3.5 for unfilled materials at ambient temperature. Filled materials were found to be less sensitive to the frictional force, whether wear took place by tearing or smearing, having values of the index n of 1.5-1.8. It is well known that for some conditions the surface of rubber becomes tacky during abrasion experiments, drum testing of tires and sometimes even for tires on the road. It has been suggested that either exudation of low molecular weight additives or degradation of the polymer to a material of low molecular weight could be responsible. Degradation might result from either thermal or mechanical stress, at high sliding speeds, such as skidding of a vehicle on locked wheels, frictional heating certainly causes degradation. However, the phenomenon of smearing is associated with conditions of mild abrasion, e.g. on smooth surfaces, and can occur even for low sliding speeds.

The most plausible mechanism of smearing appears to be oxidative consummation of scissions produced by mechanical stress, in much the same way as occurs during cold mastication of NR. Similar experimental observations to those of Gent and Pulford (Gent & Pulford, 1983) were previously obtained by Rudakov and Kuvshinski (Rudakov & Kuvshinski, 1967) for abrasion of NR and BR by a smooth indenter in air and in helium. They also gave a calculation suggesting that the rise in temperature of the rubber surface was quite inadequate to cause thermal degradation. However, this calculation ducks the possibility of local hotspots: the smaller the region of real contact, the higher is the calculated temperature rise, but we can only conjecture as to the size of the real contacts (Schallamach, 1967).

Schallamach (Schallamach, 1968) investigated the factors influencing smearing on the Akron laboratory abrader. He found that smearing could be prevented for NR tire tread compounds by carrying out abrasion in nitrogen or obviated by feeding a dust (magnesia proved most effective) into the nip between test piece and abrasive wheel. He concluded that oxidative degradation (to which he attributed smearing) affects the rate of abrasion in two distinct ways. If smearing occurs, the rate of abrasion is reduced (presumably because the "smear" acts as a lubricant). When the abrasion of a rubber is low in air, owing to smearing, its abrasion in nitrogen can become greater than in air. However, in air the less grossly degraded rubber is mechanically weakened, so that if smearing is obviated by the use of a suitable dust, the rate of abrasion is greater in air than in nitrogen. He also showed that the susceptibility of the compound to oxidative degradation can be influenced by the choice of antioxidant and other formulation details. Pulford (Pulford, 1983) studied antioxidant effects during abrasion of NR tire tread compounds by a razor blade. He reported that all compounds exhibited smearing at sufficiently low friction loads, but antioxidants reduce the critical frictional force below which smearing occurs. He found that antioxidants reduce the rate of wear for conditions in which smearing occurs but have no effect at higher severities. He considered this to be evidence of two mechanisms of wear, namely degradation at low frictional force and fracture at high frictional force. However, antioxidants also protect against fatigue crack growth, but only at low tearing energies (Lake, 1983). Thus it may not be necessary to invoke an entirely different mechanism of abrasion when smearing occurs. Instead, smearing can be seen as a complication superimposed on the general fracture mechanism of abrasion. Antioxidants can be used to, at least, partially restore the abrasion and crack growth resistance.

5. Conclusions

Abrasion resistance is the ability of a material to withstand mechanical action such as rubbing, scraping, or erosion that tends progressively to remove material from its surface. Such an ability helps to maintain the material's original appearance and structure. Numerous companies manufacture abrasion resistant products for a variety of applications, including products which can be custom fabricated to meet the needs of specific users. Abrasion resistant materials can be used for both moving and fixed parts. In vulcanized material or synthetic rubber compounds, a measure of abrasion resistance relative to a standard rubber compound under defined conditions. The properties of a vulcanized rubber can be significantly influenced by details of the compounding. Practical materials will have, in addition to the base polymer, fillers, antioxidants, crosslinking agents, accelerators etc. All of these can have an influence on the physical and chemical stability of the finished material. For example, rubber abrasion resistance can be related quantitatively to the frictional force and the crack growth characteristic of the rubber. Rigid filler particles normally increase the abrasive wear loss of the filled rubbers. A lubricant may cause a small decrease in frictional force but a dramatic decrease in abrasion. Antioxidants can be used to, at least, partially restore the abrasion and crack growth resistance because they are added to prevent ozone attack on the rubber, which can lead to the formation and growth of cracks.

6. References

Arayapranee,W.; Na-Ranong, N. & Rempel, G. L. (2005). Application of Rice Husk Ash as Fillers in the Natural Rubber Industry. *Journal of Applied Polymer Science*, Vol. 98, pp. 34–41.

Arayapranee,W. & Rempel, G. L. (2008). A comparison of the properties of rice husk ash, silica, and calcium carbonate filled 75:25 NR/EPDM blends. *Journal of Applied Polymer Science*, Vol. 110, pp. 1165–1174.

Arayapranee,W. & Rempel, G. L. (2009). Synthesis and mechanical properties of diimide-hydrogenated natural rubber vulcanizates. *Journal of Applied Polymer Science*, Vol. 114, pp. 4066–4075.

Champ, D.H. ; Southern, E. & Thomas, A.G. (1974). Fracture mechanics applied to rubber abrasion. *American Chemical Society, Division of Organic Coatings and Plastics Chemistry*, Vol. 34, No. 1, pp. 237-243.

Chandrasekaran, M. & Batchelor, A.W. (1997). In situ observation of sliding wear tests of butyl rubber in the presence of lubricants in an X-ray micro focus instrument. *Wear*, Vol. 211(1), pp. 35–43.

Dick, J. S. (2001). Vulcanizate physical properties, performance characteristics and testing, In: *Rubber Technology Compounding and Testing for Performance*, 66-67, Hanser, ISBN 1-56990-278-X, Ohio, USA.

El-Tayeb, N. S. M. & Nasir, R. Md. (2007). Effect of soft carbon black on tribology of deproteinised and polyisoprene rubbers, *Wear*, Vol. 262, pp 350-361.

Evstratov, V. V.; Reznikovski, M. M.; Smirnova, L. A. & Sakhhinovabi, N. L. (1967). The mechanism of wear of tread rubbers, in: D.I. James (Ed.), Abrasion of Rubber, Maclaren, London, pp. 45–63.

Fukahori, Y. & Yamazaki, H. (1994). Mechanism of rubber abrasion—Part 1: abrasion pattern formation in natural rubber vulcanizate. *Wear*. Vol. 171, pp. 195–202.

Gent, A. N. (1989). A hypothetical mechanism for rubber abrasion. *Rubber Chemistry & Technology*, Vol. 62, pp. 750-756.

Gent, A. N. & Pulford, C. T. R. (1983). Mechanisms of rubber abrasion. *Journal of Applied Polymer Science*, Vol. 28, pp. 943-960.

Hong, C. K.; Kim, H. ; Ryu, C. ; Nah C. ; Huh, Y.-il & Kaang, S. (2007). Effect of paticle size and structure of carbon blacks on the abrasion of filled elastomer compounds, *Journal of Materials Science*, Vol 42, pp. 8391-8399.

Jurkowska, B.; Jurkowski, B.; Nadolny, K.; Krasnov, A. P.; Studniev, Y. N.; Pesetskii. S. S.; Koval, V. N.; Pinchuk, L. S. & Olkhov, Y. A. (2006). Influence of fluorine-containing lubricant on properties of NR/BR rubber. *European Polymer Journal*, Vol. 42, pp. 1676-1687.

Lake, G. J. (1983). Aspects of fatique and fracture of rubber. *Progress of Rubber Technology*, Vol. 45, pp. 89-143.

Liang, H.; Fukahori, Y.; Thomas, A. G. & Busfield, J. J. C. (2010). The steady state abrasion of rubber: Why are the weakest rubber compounds so good in abrasion? *Wear*, Vol. 268, pp. 756-762.

Muhr. A. H. & Roberts, A. D. (1992). Rubber abrasion and wear. *Wear*, Vol. 158, pp. 213-228.

Pandey, K. N.; Setua, D. K. & Mathur, G. N. (2003). Material behaviour fracture topography of rubber surfaces: an SEM study, *Polymer Testing*, Vol. 22, pp. 353–359.

Pulford, C. T. R. (1983). Antioxidant effects during blade abrasion of natural rubber. *Journal of Applied Polymer Science*, Vol. 28, pp. 709-713.

Ratner, S. B.; Farberova, I. I.; Radyukevich, O. V. & Lu're, E. G. (1967). Connection between wear resistance of plastics and other mechanical properties, In: *Abrasion of Rubber*, James, D. I. (Ed.), Macaren, London, pp. 145-154.

Roberts, A. D. (1988). Rubber adhesion at high rolling speeds. Journal of Natural Rubber Research. Vol 3, pp 239.

Rudakov, A. P. & Kuvshinski, E. V. (1967). Abrasion of rubber by a smooth indentor, In: *Abrasion of Rubber*, James, D. I (Ed.), Maclaren, London, pp. 36-44.

Sae-oui, P.; Rakdee, C. & Thanmathorn, P. (2002). Use of rice husk ash as filler in natural rubber vulcanizates: In comparison with other commercial fillers. *Journal of Applied Polymer Science*, Vol. 83, pp. 2485-2493.

Schallamach, A. (1957/58). Friction and abrasion, *Wear*, Vol. 1, pp. 384-417.

Schallamach, A. (1967). A note on the frictional temperature rise of tyres, *Journal of the Institution of the Rubber Industry*, Vol. 1, pp. 40-42.

Schallamach, A. (1968). Abrasion, fatigue, and smearing of rubber. *Journal of Applied Polymer Science*, Vol. 12, pp 281-293.

Tabsan, N.; Wirasate, S. & Suchiva, K. (2010). Abrasion behavior of layered silicate reinforced natural rubber. *Wear*, Vol. 269, pp. 394-404.

Thomas, A. G. (1958). Rupture of rubber. V. cut growth in natural rubber vulcanizates. *Journal of Polymer Science*. Vol. 31, pp. 467-480.

Thomas, A. G. (1974). Factors influencing the strength of rubbers. *Journal of Polymer Science*, Vol. 48. pp. 145-157.

Uchiyama, Y. & Ishino, Y. (1992). Pattern abrasion mechanism of rubber. *Wear*. Vol. 158, pp. 141–155.

Yang, A. C.-M.; Ayala, J. E. & Campbell Scott, J. (1991). Abrasive wear in filled elastomers. *Journal of Materials Science*, Vol. 26, pp. 5823-5837.

9

Effect of Abrasive Size on Wear

J. J. Coronado
Research Group of Fatigue and Surfaces,
Mechanical Engineering School, Universidad del Valle, Cali
Colombia

1. Introduction

Due to their high wear resistance, metallic materials are usually used in some mechanical components used for extraction, processing and transportation in mining and agriculture industries. These mechanical components work in contact with hard abrasive particles of different sizes and morphologies, and, in many cases, in highly corrosive environments. As in all other forms of wear, the intensity of abrasive wear depends on the configuration of the wear system, the force applied, and the microstructural properties of materials. Abrasive wear, however, has the peculiarity of being strongly influenced by the characteristics of the abrasive particles, such as hardness, morphology and size.

There are two major groups in which abrasive wear is classified: two body-abrasive wear and three-body abrasive wear (Rabinowicz, 1961; Misra; Finnie, 1980). Two-body abrasive wear occurs when abrasive particles are fixed to one body while the second one slides over it, scratching or removing material, as, for example, in the pin-on-disc test. In three-body abrasive wear particles are free to roll and, as a result, they do not remove material from the body all the time that they are in contact as, for example, in the rubber-wheel equipment.

In abrasive wear, the material is damaged or removed by several mechanisms. In Figure 1 three mechanisms of abrasive wear are shown by scanning electron microscopy (SEM): microploughing, wedge formation, and microcutting (Hokkirigawa; Kato, 1988; Kato, 1990).

The mechanism of microploughing causes displacement of material, resulting in the formation of side edges. In the microcutting the material is removed in the form of microchips, this mechanism operates similarly to a cutting tool. The mechanisms of microploughing and microcutting are related to moderate and severe wear, respectively. Wedge formation is associated with the transition between microploughing and microcutting (Kayaba et al. 1986; Hokkirigawa *et al.*, 1987).

2. Abrasive morphology

Literature reports that material can be detached from the surface by microcutting when the attack angle of the abrasive particles is higher than the critical attack angle (Mulhearn; Samuels, 1962; Sedriks; Mulhearn, 1963; Sedriks; Mulhearn, 1964). A gradual transition from microploughing to microcutting occurs when the attack angle increases (Zum Gahr, 1987). This is shown in Figure 2. The attack angle (α) is given by the angle formed between the abrasive surface and the material surface.

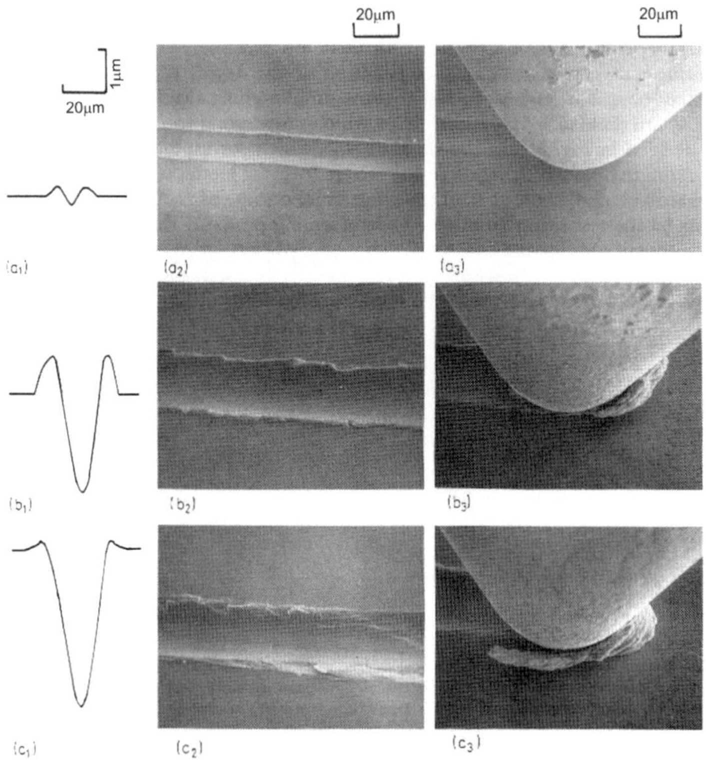

Fig. 1. Wear micromechanisms observed by SEM: (a) microploughing, (b) wedge formation (c) microcutting (Hokkirigawa; Kato, 1988)

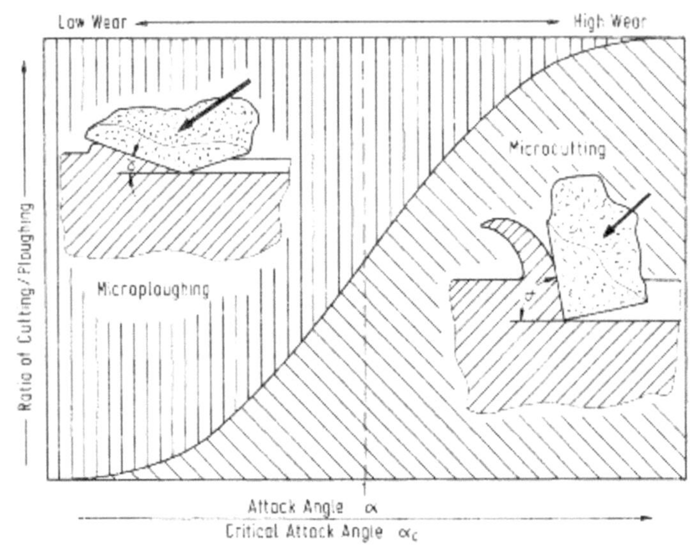

Fig. 2. Ratio of microcutting to microploughing as a function of the ratio of the attack angle to the critical attack angle (α_c) (Zum Gahr, 1987)

A practical example is the aluminum alloy sheets used in trucks for transporting granular material, which show damage caused by the abrasion of sand and gravel particles. An analysis of granular material show abrasive particles of different sizes and morphologies: large particles tend to have a round shape (figure 3a), while small particles have greater angularity (figure 3b) (Mezlini *et al.* 2005).

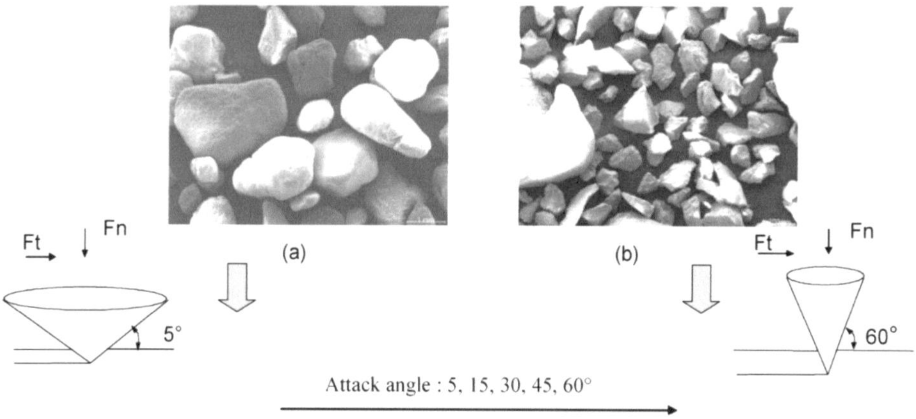

Fig. 3. SEM observations of transported granular material: (a) rounded particles, (b) sharp particles. (Mezlini *et al.*, 2005)

Mezlini *et al.* (2005) performed scratch tests on 5xxx aluminum alloy to study the effect of the attack angle (geometry of the particle) on abrasive wear micromechanisms. The authors used rigid cones with different attack angles (Figure 3). When an indenter with an attack angle of 30° is used, the material is moved to the edges of the scratch, so the material is accumulated along the edges and in front of the indenter without loss of material (microploughing mechanism), this is show in Figure 4(a). When using an indenter with an attack angle of 60°, a transition in wear micromechanisms from microploughing to microcutting is produced with chip formation in front of the indenter (figure 4(b)).

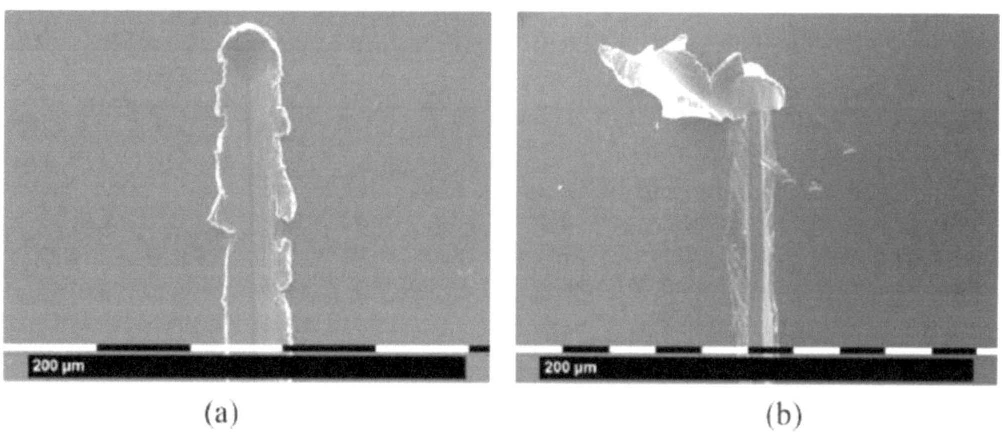

Fig. 4. SEM observations of aluminum alloy scratches for an attack angle: (a) 30° and (b) 60° (Mezlini *et al.*, 2005)

3. Abrasive size

The literature reports that the abrasive size has a linear relationship with mass loss for small abrasives. The effect of the abrasive size on the wear rate has been studied for homogeneous materials (Anvient, 1960; Rabinowicz; Dunn, 1961; Goddard; Wilman, 1962; Rabinowicz; Mutis, 1965; Nathan; Jones, 1966; Larsen-Badse, 1968a; Larsen-Badse, 1968b; Samuels, 1971; Date; Malkin, 1976; Sin *et al.*, 1979; Misra; Finnie, 1981a; Misra; Finnie, 1981b; Sasada *et al.*, 1984; Jacobson *et al.*, 1988; Costa et. al., 1997; Gahlin; Jacobson, 1999; Sevim; Eryurek, 2006). For small abrasives, the wear rate increases proportionally with the increase in the abrasive particle size until it reaches the critical particle size (CPS). After reaching the critical particle size, the wear rate changes. Figure 5 summarizes the three behaviors described in the literature.

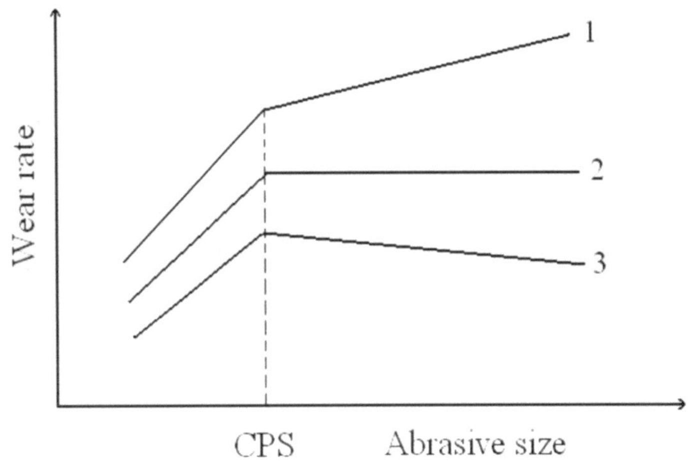

Fig. 5. Schematic representation showing the typical curves of wear vs. abrasive grain size (Coronado; Sinatora, 2011a)

After the CPS, three phenomena can occur: the wear rate can increase at a lower rate (curve 1), it can become constant, independent of further abrasive size increases (curve 2), or it can exhibit a decreasing rate (curve 3). There are many hypotheses to explain this phenomenon; however, there is still no explanation generally accepted by the entire scientific community. The phenomenon of CPS occurs in two-body abrasive wear, three-body abrasive wear, erosion and machining processes. Because of the importance of the effect of abrasive size, both in tribology and in manufacturing processes, a detailed discussion of the literature is presented chronologically in this chapter.

Anvient *et al.* (1960) conducted tests of abrasive wear in pure metals: Ag, Cu, Pt, Fe, Mo and W. The results are shown in Figure 6. The authors reported that the wear rate increased with the increase in the abrasive size in the range of 5 to 70 μm and was independent of the wear rate within the abrasive size ranges of 70 to 140 μm.

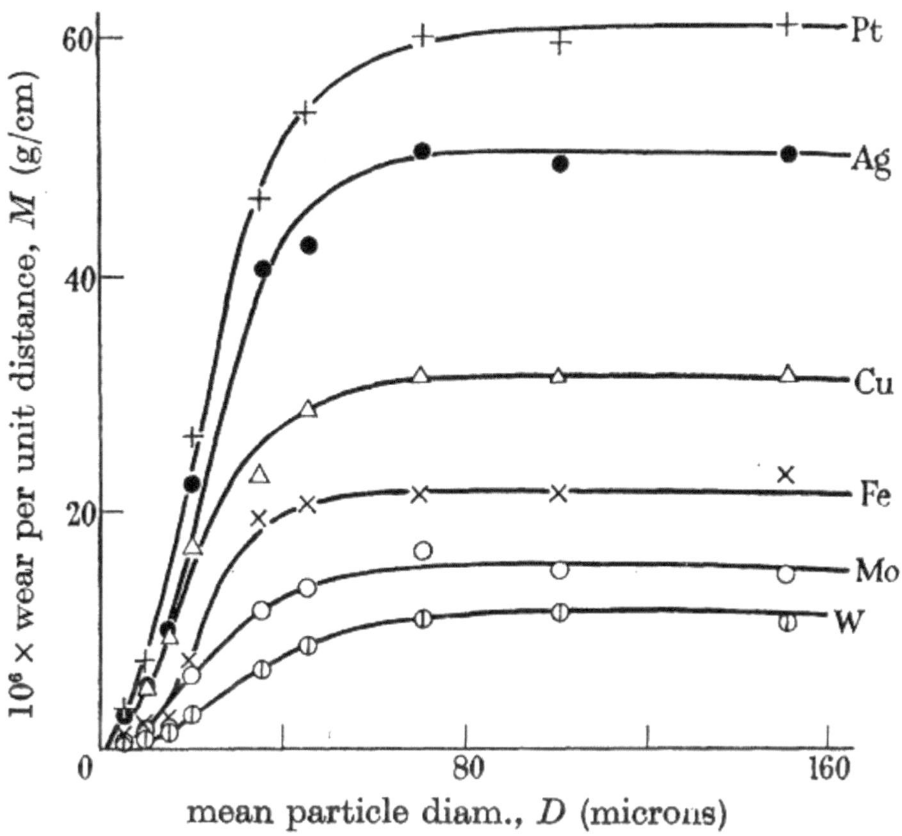

Fig. 6. Relationship between the wear rate and mean abrasive size (Anvient *et al.*, 1960)

Anvient *et al.* (1960) and Goddard and Wilman (1962) proposed that the CPS is controlled by clogging of the smaller sized abrasives. Due to the clogging is not possible in three-body abrasive and erosion, this explanation cannot explain de CPS effect (Misra; Finnie, 1981a).

Nathan and Jones (1966) conducted wear tests in copper, aluminum, brass and steel using a two-body abrasive wear equipment. The authors studied the correlation between variation of volume of wear, variation of diameter of abrasive particles of SiC (35 to 710 μm), load (0.5 - 6 kg), speed of abrasion (0.032 - 2.5 m/sec), and distance traveled (1.5 - 6 m). The results presented in Figure 7 indicate that the volume of material removed increases linearly with the size of the abrasive particles up to 70 μm. Between 70 μm and 150 μm, the gradient continuously decreases, and above 150 μm, it presents a linear relationship at a lower rate.

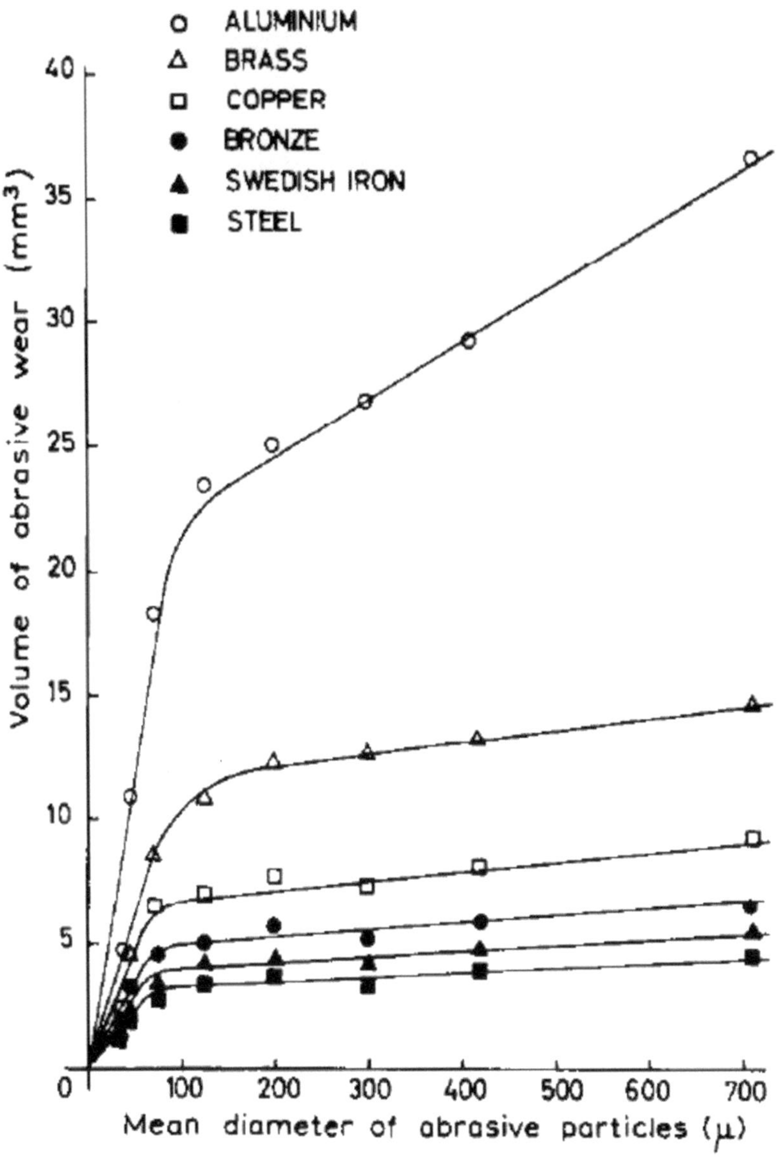

Fig. 7. Relationship between the volume of abrasive wear and the mean diameter of abrasive particles (Nathan; Jones, 1966)

Larsen-Badse (1968a) carried out two-body abrasive wear tests on copper using SiC as abrasive. The author found that the wear rate increases rapidly until it reaches the CPS. The value of CPS was in the range of 40 μm to 80 μm. Above the CPS, two things happened: the wear rate was constant for low loads and the wear rate decreased for high values of load. The results are shown in Figure 8.

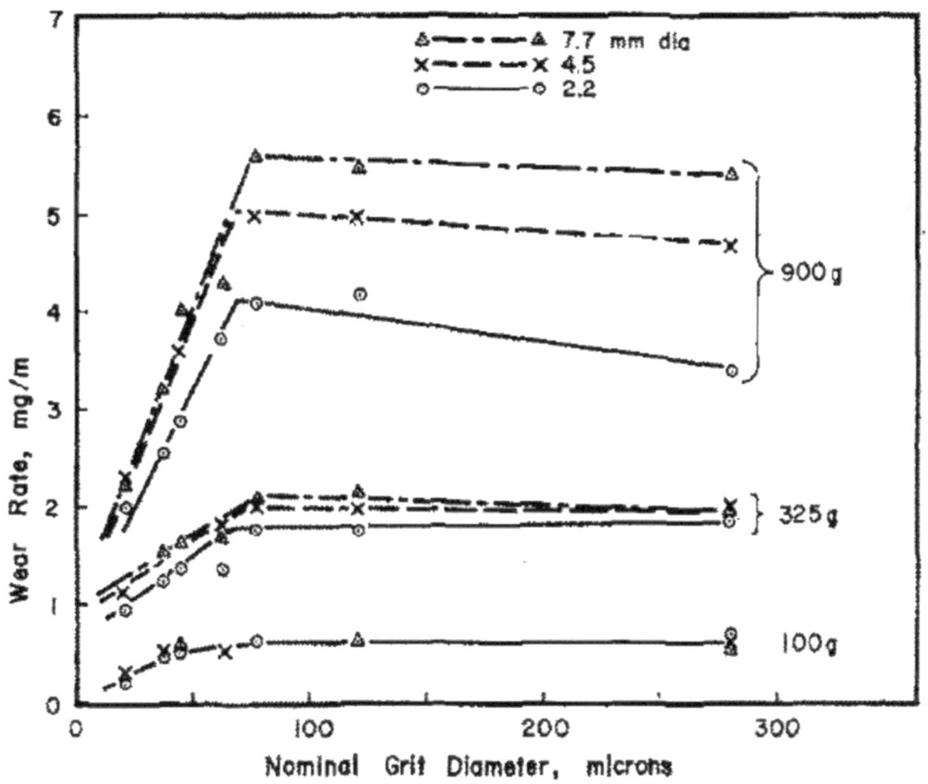

Fig. 8. Effect of grit size on wear rates (Larsen-Badse, 1968a)

Damage of smaller abrasives has been proposed by Larsen-Badse (1968a) to explain the effect of abrasive size. The explanation of the damage of smaller abrasive, however, was not corroborated, for when the abrasive size increases, the probability of finding more defects increases.

Date and Malking (1976) realized abrasion tests on AISI 1090 steel using alumina as abrasive (# 320, # 240, # 150, # 80 and # 36). The results are shown in Figure 9.

These authors performed an extensive study on the abrasives after wear using SEM and found that the largest abrasive had more damage. Date and Malking (1976) also found a deposit of material covering the surface of the abrasive below the CPS. A similar but less intense effect was present in bigger abrasives. Microchips resulting from abrasion and debris resulting from adhesion in the interstices of the smaller abrasives were also responsible for the decrease in wear rate with the decrease of abrasive size (Date; Malking, 1976). The tests were not performed on unused abrasives. Some complications observed in some of the previous work, such as clogging, could be avoided if the tests were conducted on unused abrasive.

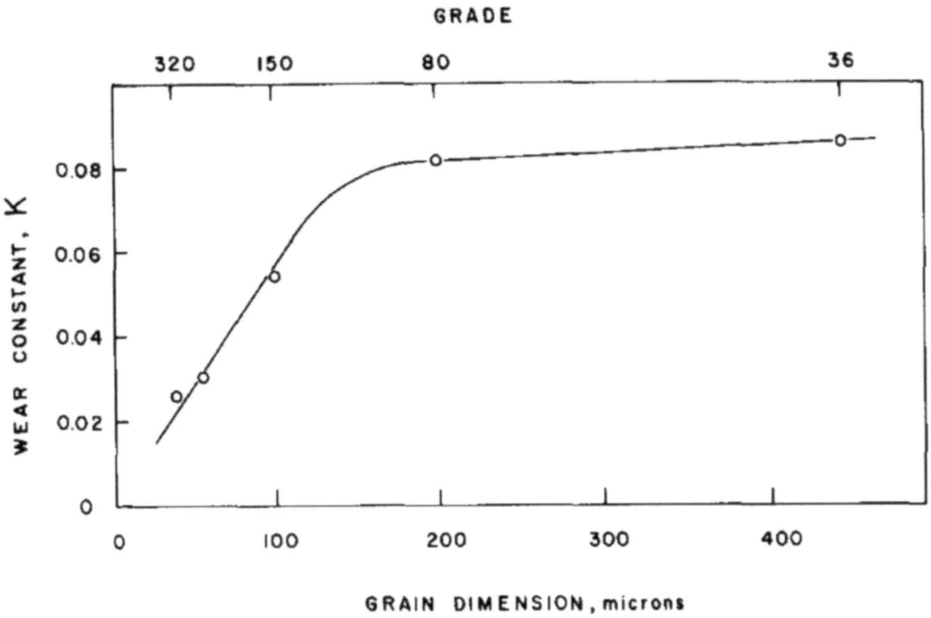

Fig. 9. Relationship between wear coefficient and abrasive size (Date; Malking, 1976)

Sin et al. (1979) conducted tests using pin-on-disc equipment in spiral trajectory using SiC abrasive. The materials tested were PMMA (polymethylmethacrylate), pure nickel and AISI 1095 steel. The results shown in Figure 10 indicate that as the abrasive size increases, the wear coefficient increases rapidly until it reaches the CPS. Above the CPS, the wear coefficient is independent of the size of the abrasive. The CPS was approximately 80 µm for the materials tested.

Sin *et al.* (1979) proposed that the effect of size of the abrasive is due to the round edges of smaller abrasive grains. Misra and Finnie (1981a) demonstrated, however, that abrasive grains of reduced sizes are more pointed. Sin *et al.* (1979) also reported that smaller abrasives with rounded ends produce more microcutting than microploughing. However, Misra and Finnie (1981a) observed the wear surfaces with SEM and found little increase in microploughing with the decrease in the abrasive size.

Misra and Finnie (1981a) carried out tests on copper using two-body abrasive wear, three-body abrasive wear and erosion. The results showed that when the abrasive SiC size is over 100 µm, the wear rate is little affected by the increase in abrasive size. Figure 11 shows that for abrasive sizes lower than 100 µm, the wear rate decreases. The wear process becomes less efficient as the particle size decreases below 100 µm. An analysis of the curves, however, shows that only two abrasive sizes larger than 100 µm were used. The authors propose that a shallow layer near the worn surface shows more flow stress than that of the bulk material. The explanation of a shallow layer was first proposed by Kramer and Demer (1961).

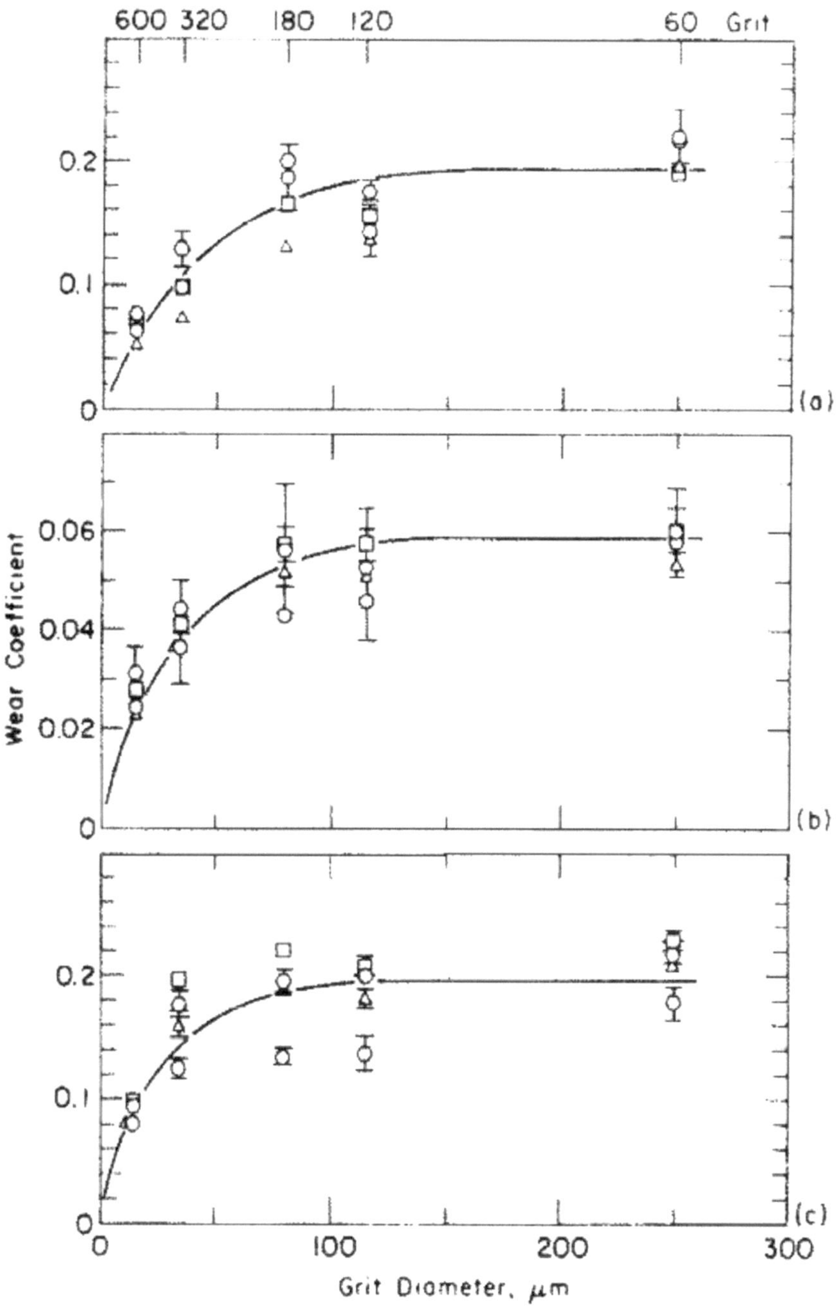

Fig. 10. Wear coefficient vs. abrasive grit diameter for different normal loads: (a) PMMA, (b) nickel and (c) AISI 1095 steel (Sin *et al.*; 1979)

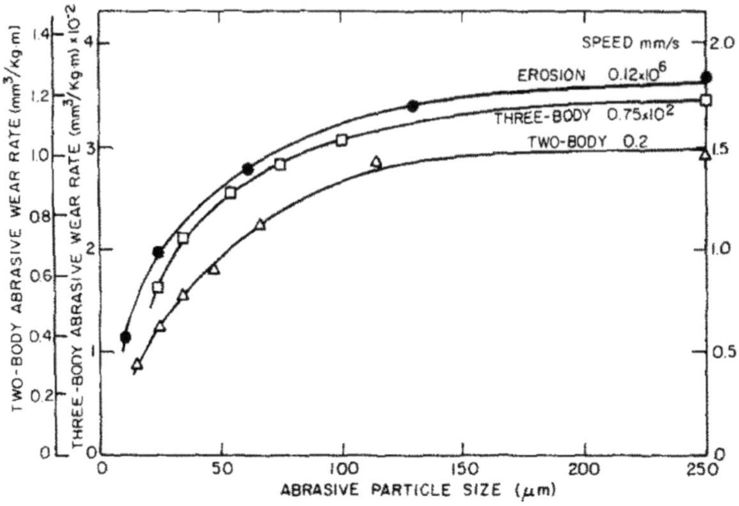

Fig. 11. Wear rate as a function of particle size for copper samples under erosion, two-body abrasion and three-body abrasion (Misra; Finnie, 1981a)

Previous research, however, was made by Goodwin *et al.* (1969). They studied the effect of abrasive size on erosion in steel using different speeds and impact angles. The authors found that above the CPS the erosion was not influenced by the size of the abrasive and this value (CPS) increased with the increase of impact velocity.

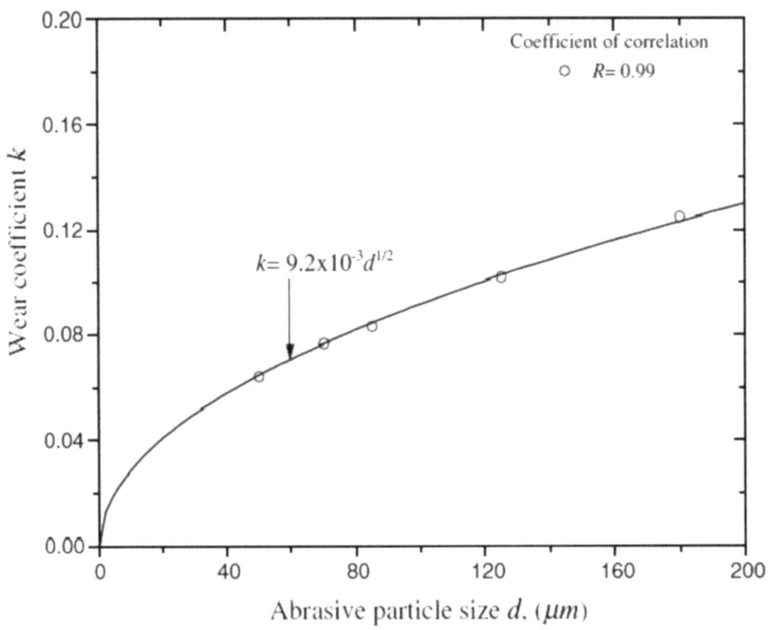

Fig. 12. Variations of wear coefficient *k* of non-heat-treated steels versus abrasive particle size *d* (Sevim; Eryurek; 2006)

In a more recent work, Sevim and Eryurek (2006) conducted tests on steels using alumina abrasives with sizes between 50 and 180 μm. Figure 12 shows that there is a parabolic relationship between the wear coefficient (*k*) and the size of the abrasive (*d*). The coefficient *k* is not constant with the increase of *d*. The slope, however, decreases with the increase of *d*. A more detailed analysis of the curve, however, indicates that a linear relationship can be used between *k* and *d* with a high correlation coefficient. The authors show that the effect of reducing the size of the abrasive grain in the severe regimen results in decreases in mass loss of about 20% to 40%.

In hard second phase materials the effect of the abrasive size on wear rate has been focused on the white cast irons with high chromium and alloy development (Santana; De Mello, 1993; Pintaúde et al., 2001; Dogan et al., 2001; Bernardes, 2005; Dogan *et al.*, 2006).

4. Recent works

In recent researches the effect of abrasive size on metallic materials using two-body configuration was investigated and the most relevant results are discussed in this chapter. In a first series of experiments the samples of mottle cast iron were quenched and tempered in temperatures ranging from 300°C to 600°C, forming different percentages of retained austenite (RA). For small abrasive particles, the wear mass loss increased linearly with the increase of particle size. However, for higher abrasive sizes the mass loss increased much more slowly (Figure 13). For lower abrasive sizes the main wear mechanism was microcutting. For higher abrasive sizes, the main wear mechanism was microploughing (Coronado; Sinatora, 2009).

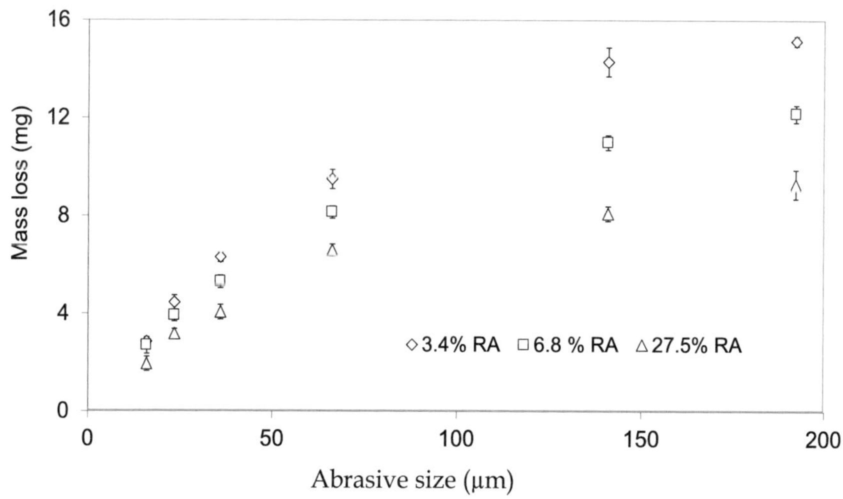

Fig. 13. Relationship between mass loss and abrasive sizes (Coronado; Sinatora, 2009)

In a second series of experiments, white cast iron with M_3C carbide with austenitic and martensitic matrix were tested (Coronado; Sinatora, 2011a). The alumina abrasives of lower size are characterized by sharp tips and the alumina abrasives of greater size are characterized by rounded edges and polyhedral shapes. The results show that the mass loss for cast irons with austenitic and martensitic matrices increases linearly with the increase of particle size, until reaching the critical particle size. After that, the rate of mass loss of the cast iron with austenitic matrix diminishes to a lower linear rate, and for cast irons with martensitic matrix the curve of mass loss is non-linear and flattens at the critical particle size. It becomes, then, constant, independent of additional size increases (figure 14). The abrasive paper in contact with the iron of both austenitic and martensitic matrices presents fine continuous microchips produced by microcutting before reaching critical particle size, and after that it presents deformed discontinuous microchips produced by microploughing (figures 15 and 16).

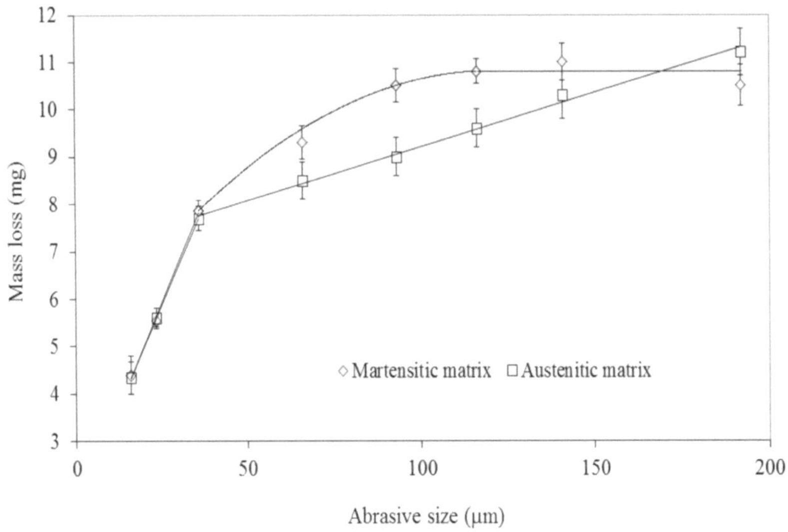

Fig. 14. Comparison between the cast iron with austenitic and martensitic matrix (Coronado; Sinatora, 2011a)

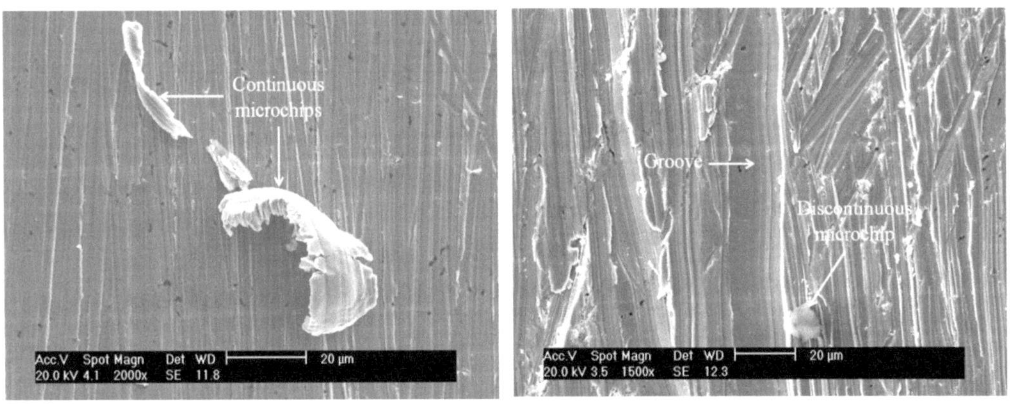

(a) (b)

Fig. 15. SEM of the wear surface for WCI with austenitic matrix (a) 23.6 μm and (b) 141 μm (Coronado; Sinatora, 2011a)

(a) (b)

Fig. 16. SEM of the abrasive paper against cast iron with austenitic matrix and abrasive size of: (a) 23.6 µm and (b) 116 µm (Coronado; Sinatora, 2011a)

In a third series of experiments, aluminum and AISI 1045 steel were tested (Coronado; Sinatora, 2011b). The first (FCC structure) showed a behavior similar to that observed in the white cast iron with austenitic matrix, and the latter showed a behavior similar to that observed in white cast iron with martensitic matrix (Figure 17). Both aluminum and AISI 1045 steel show similar changes in microchip morphology and in wear micromechanisms, something that had been observed before in materials with a hard second phase (Figures 18 and 19).

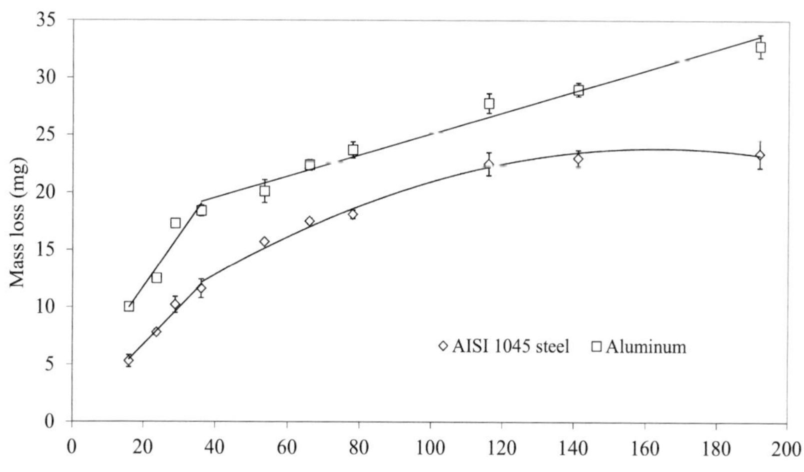

Fig. 17. Comparison between the AISI 1045 steel and the aluminum alloy (Coronado; Sinatora, 2011b)

(a) (b)

Fig. 18. SEM of the abrasive paper against AISI 1045 steel and abrasive size of: (a) 23.6 µm and (b) 141 µm (Coronado; Sinatora, 2011b)

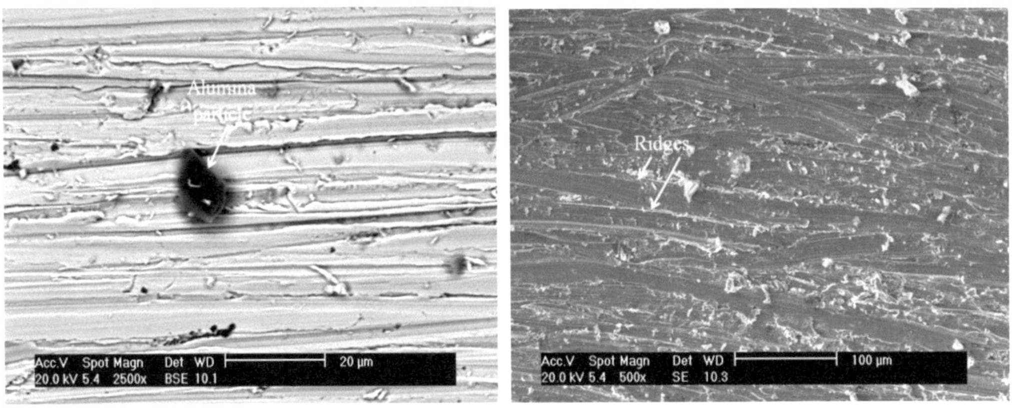

(a) (b)

Fig. 19. SEM of the wear surface of AISI 1045 steel using alumina of: (a) 23.5 µm and (b) 141 µm (Coronado; Sinatora, 2011b)

Finally, in a fourth series of experiments, gray cast iron was tested in order to demonstrate the relationship between the abrasive wear micromechanisms and the type of microchips, before and after achieving critical abrasive size (Coronado; Sinatora, 2011b). The gray cast iron did not show a transition in the curve of abrasive size against mass loss (figure 20). The morphology of the microchips was similar (discontinuous) for the different sizes of abrasive. However, smaller abrasive sizes - some thin continuous microchips - were formed (figure 21). The main abrasive wear micromechanism was microcutting for the different abrasives sizes tested (Figure 22). This, therefore, shows that the critical abrasive size is related to the wear micromechanisms and the microchip morphology.

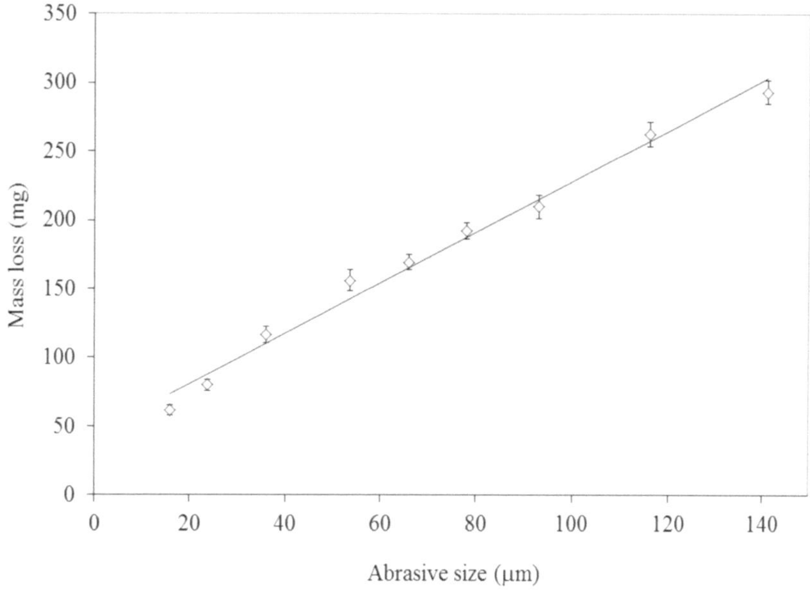

Fig. 20. Relationship between mass loss and abrasive size for gray cast iron (Coronado; Sinatora, 2011b)

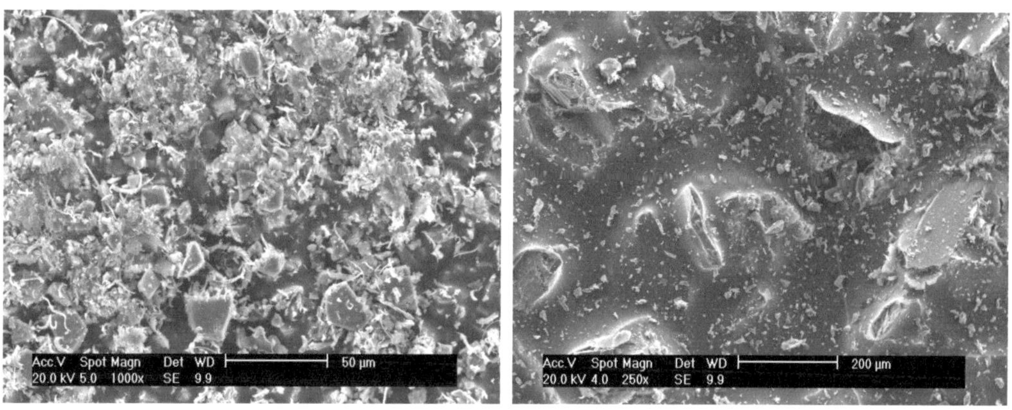

(a) (b)

Fig. 21. SEM of the abrasive paper against gray cast iron and abrasive size of (a) 16 μm and (b) 116 μm (Coronado; Sinatora, 2011b)

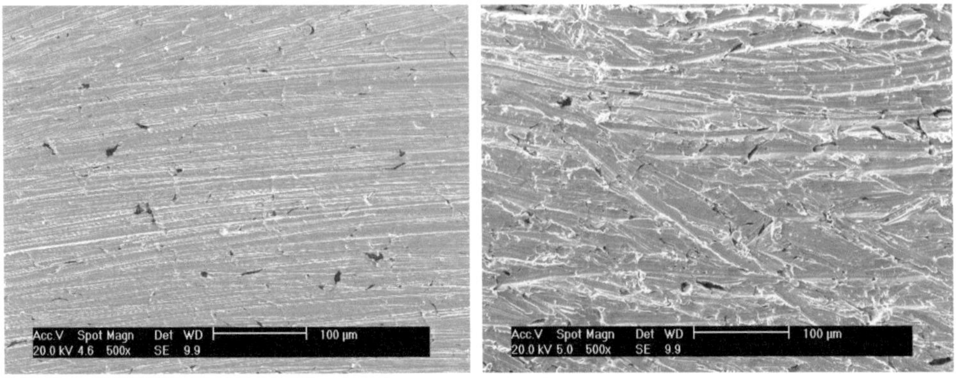

(a) (b)

Fig. 22. SEM of the wear surface of gray cast iron using alumina of: (a) 16 µm and (b) 116 µm (Coronado; Sinatora, 2011b)

5. Conclusions and future trends

For metallic materials and using two-body abrasive configuration, the following general conclusion can be obtained: at lower abrasive sizes the sharp tips cut the material with lower penetration producing continuous undeformed microchips. However, at some critical abrasive sizes, the prevalent wear mechanism changes from microcutting to microploughing, producing discontinuous deformed microchips. With abrasives of bigger sizes with round edges and polyhedral shapes, the microploughing component decreases the wear rate because the plastic deformation becomes more important than the microcutting action (Coronado; Sinatora, 2011a,b).

Although recent researches show an adequate explanation to the phenomenon of critical abrasive size, there can be no assurance that these results can be generalized for all abrasive wear settings. In the future, the effect of grain size in three-body abrasive wear and erosion should be studied, in order to know whether the results found in the two-body abrasive wear in metals are also valid for three-body abrasive wear and erosion.

Nowadays, there is a significant potential growth in the use of ceramics and polymers with the purpose of protection of components subjected to wear, but studies in the literature do not mention abrasive size effect in these materials, with some exceptions. The study of this phenomenon in ceramic materials and polymers in two-body abrasive tests is an area of research and practical interest yet to be developed.

One thing the development of research in this area has demonstrated is that in the analysis of any engineering materials, a good approach to follow is to observe the effect of abrasive size in composite materials using two-body abrasive wear configuration in correlation with the wear micromechanisms and microchips formed, comparing results with those from the materials previously researched.

6. References

Avient, W. E., Goddard, J. Wilman, H. An experimental study of friction and wear during abrasion of metals. Proceedings of the Royal Society of London. Series A. Mathematical and Physical Sciences 258, 1960, p.159-180.

Bernardes, G. F. Desgaste abrasivo de um ferro fundido branco multicomponente. Dissertação de Mestrado Engenharia Mecânica. Universidade de São Paulo. 2005.

Coronado, J. J. Sinatora, A. Particle Size Effect on Abrasion Resistance of Mottled Cast Iron with Different Retained Austenite Contents. Wear 267, 2009, p. 2077-2082.

Coronado, J. J. Sinatora, A. Effect of abrasive size on wear of metallic materials and its relationship with microchips morphology and wear micromechanisms: Part 1. Wear 271, 2011a, p. 1794- 1803.

Coronado, J. J. Sinatora, A. Effect of abrasive size on wear of metallic materials and its relationship with microchips morphology and wear micromechanisms: Part 2. Wear 271, 2011b, p. 1804- 1812.

Costa, H. L. Pandolfelli, V. C. De Mello, J. D. B. On the abrasive wear of zirconias, Wear 203-204, 1997, p.626-636

Date, S. W Malkin, S. Effects of grit size on abrasion with coated abrasives, Wear 40, 1976, p.223–235.

Dogan, Ö. N. Hawk, J. A. Tylczak, J. H. Wear of cast chromium steels with TiC reinforcement, Wear 250, 2001, p.462–469.

Dogan, Ö. N. Hawk, J. A. Schrems, K. K. TiC-Reinforced Cast Cr Steels, Journal of Materials Engineering and Performance 15, 2006 p.320–327.

Gahlin, R. Jacobson S. The particle size effect in abrasion studied by controlled abrasive, Wear 224, 1999, p.118–125.

Goddard, J. Wilman, H. A theory of friction and wear during the abrasion of metals, Wear 5, 1962, p.114–135.

Hokkirigawa, K. Kato, K. An experimental and theoretical investigation of ploughing, cutting and wedge formation during abrasive wear. Tribology International 21, 1988, p.51–57.

Hokkirigawa, K. Kato, K. Li, Z. Z. The effect of hardness on the transition of the abrasive wear mechanism of steels, Wear 123, 1987, p.241–51.

Jacobson, S. Wallen, P. Hogmark, S. Fundamental aspects of abrasive wear studied by a new numerical simulation model, Wear 123, 1988, p.207–223.

Kato, K. Wear mode transitions, Scripta Metallurgica 24, 1990, p.815–820.

Kayaba, T. Hokkirigawa, K. Kato, K. Analysis of the abrasive wear mechanism by successive observations of wear processes in a scanning electron microscope, Wear 110, 1986, p.419–30.

Larsen-Badse, J. Influence of grit diameter and specimen size on wear during sliding abrasion, Wear 12, 1968a, p.35–53.

Larsen-Badse, J. Abrasion resistance of some S.A.P. type alloys at room temperature, Wear 12, 1968b, p.357-368.

Mezlini, S. Zidi, M. Arfa, H. Tkaya, M. B. Kapsa, P. Experimental, numerical and analytical studies of abrasive wear: correlation between wear mechanisms and friction coefficient, C. R. Mecanique 333, 2005, p. 830–837.

Misra, A. Finnie, I. A classification of three-body abrasive wear and design of a new tester, Wear 60, 1980, p.111-121.

Misra, A. Finnie, I. On the size effect in abrasive and erosive wear, Wear 65, 1981a, p.359–373.

Misra, A. Finnie, I. Some observations on two-body abrasive wear, Wear 68 1981b, p.41–56.

Mulhearn, T. 0. Samuels, L. E. The abrasion of metals: a model of the process, Wear 5, 1962, p.478-498.

Nathan, G. W. Jones, J. D. The empirical relationship between abrasive wear and applied conditions, Wear 9, 1966, p.300–309.

Pintaúde, G. Tschiptschin, A.P. Tanaka, D.K. Sinatora, A. The particle size effect on abrasive wear of high-chromium white cast iron mill balls, Wear 250, 2001, p.66–70.

Rabinowicz, E. Dunn, L. A. Russell P. G. A study of abrasive wear under three-body conditions, Wear 4, 1961, p.345–355.

Rabinowicz, E. Mutis, A. Effect of abrasive particle size on wear, Wear 8, 1965, p.381–390.

Santana, S. A. De Mello, J. D. B. Influencia da morfologia de carbonetos M_7C_3 no comportamento em abrasão de ferros fundidos brancos de alto cromo. Proc. 48 th Congresso Anual da ABM. Vol. 1. Associação Brasileira de Metais. Rio de Janeiro. 1993 p. 457-476.

Samuels, L.E. Metallographic polishing by mechanical methods, Second edition, Sir Isaac Pitman & Sons Ltd. Melbourne & London, 1971.

Sedriks, A. J. Mulhearn, T. O. Mechanics of cutting and rubbing in simulated abrasive processes, Wear 6, 1963, p.457-466.

Sedriks, A. J. Mulhearn T. O. The effect of work-hardening on the mechanics of cutting in simulated abrasive processes, Wear 7, 1964, p.451-459.

Sevim, I. Eryurek, I. B. Effect of abrasive particle size on wear resistance in steels, Materials and Design 27, 2006, p.173–181.

Sin, H. Saka, N. Suh, N. P. Abrasive wear mechanism and the grit size effect, Wear 55, 1979, p.163–190.

Zum Gahr K. H. Microstructure and wear of materials. Elsevier, 1987.

10

Abrasion Resistance of Cement-Based Composites

Wei-Ting Lin[1,2] and An Cheng[1]
[1]*Department of Civil Engineering, National Ilan University, Ilan,*
[2]*Institute of Nuclear Energy Research, Atomic Energy Council, Taoyuan,*
Taiwan

1. Introduction

1.1 Background

Cement-based composites are among the most widely-used construction materials due to their low cost, high compressive strength, high durability, versatility, and easy-handling. Unfortunately, cement-based composites are intrinsically porous and may deteriorate as a result of exposure to harsh environments or poor construction quality. Over the last few decades, most research has focused on the strength characteristics of concrete, with far less attention paid to material parameters influencing durability. The deterioration of these materials often results in severe damage to concrete structures such as cracking, delaminating, spalling, and even fractures. This kind of damage is generally not detected until it has reached a critical level, at which point, rust is visible on the rebar and evidence of cover concrete deterioration can be found throughout the entire structure, as shown in Fig. 1-1. The degradation of cement-based composites is considered a key factor in the durability of structures and a major concern for civil engineers.

(a) Observation in July 2007 (b) Observation in April 2008

Fig. 1.1. Corrosion damage in a concrete wall

Degradation of concrete composites is classified as physical or chemical. Physical degradation can be divided into frost action, cracking, thermal cracking due to shrinkage, fire damage, and surface abrasion. Chemical degradation can be divided into corrosion of the rebar, attack by sulfates or acids, and degradation resulting from alkali aggregates. Degradation is often evaluated in terms of permeation, which is directly linked to the movement of aggressive agents into and out of cement-based composites. Expansion causes internal stress resulting in cracking or scaling. Chemical degradation alters hydration products and often leads to dissolution or leaching. Several forms of composite degradation are the result of combined physical and chemical attack. The thickness and quality of the cover or surface layer are important factors, which determine the ability of a material to resist physical and chemical attack. Measures that have proven effective in minimizing the problems of durability include the creation of less permeable composites or denser pastes, which inhibit the propagation of cracks, and provide a cover of adequate thickness [1-3]. The density of paste can be enhanced by lowering the water cementitious ratio or through the addition of supplementary cementitious materials (SCMs). SCMs such as silica fume, fly ash, and ground granulated blast furnace slag (ggbs), are commonly used to replace a portion of the cement in cement-based composites to improve the quality and/or durability of cement-based composites. Crack inhibition can be increased through the addition of fibers [4-6].

The influence of abrasion in cement-based composites is slight and weight loss or erosion resulting from abrasion is gradual; although the problem can be exacerbated through exposure to harsh environments (Fig. 1-2). A great deal of research has been dedicated to cracking behavior and the durability of cement-based composites; however, little effort has gone into the issue of abrasion resistance. A number of structures, such as existing dams, concrete nuclear structures, and underground storage or radioactive waste disposal containers, are in constant contact with water or abrasives for extended periods of time. Such exposure increases the risk of reduced service life. Indeed, further investigation of the abrasion resistance of cement-based composites could provide considerable benefits in ensuring that structures continue to serve their intended function.

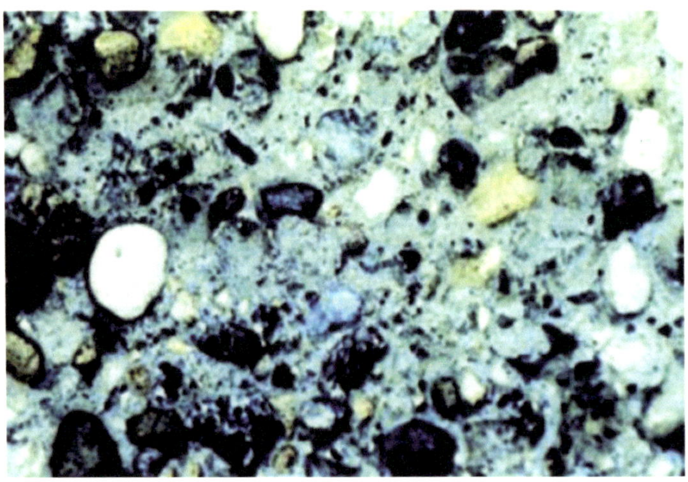

Fig. 1.2. Concrete disintegration due to abrasion [7]

1.2 Objectives

The aim of this study was to deepen our understanding of abrasion resistance in cement-based composites, through the evaluation of testing methods and material variables. The results of this study are presented in three parts:

1. Cement-based composites containing supplementary cementitious materials;
2. Cement-based composites containing fiber-reinforced materials;
3. Evaluation of indices of various testing methods.

The objectives of this study are outlined as follows:

4. Evaluate the effects of fibers and SCMs on the abrasion resistance of cement-based composites;
5. Compare the difference between the ASTM C418 (sand blast abrasion method) and the ASTM C131 (modified version, Los Angeles abrasion method).

The study includes four chapters. Chapter 1 provides an introduction to the durability and abrasion of concrete containing composites, including background information and the objectives of this study. Chapter 2 describes the testing program including ASTM C418, ASTM C779, ASTM C1138, and ASTM C131. Material variables including SCMs and fibers added to cement-based composites are also explained. Chapter 3 presents the results and discussion of abrasion resistance using various abrasion methods. Finally, Chapter 4 summarizes the main conclusions and provides recommendations for further study.

2. Literature review

2.1 Supplementary cementitious materials

In recent years, SCMs have often been used to replace a portion of the cement in cement-based composites, with the aim of improving mechanical properties or durability. The classification and function of the mineral admixtures are illustrated in Fig. 2-1.

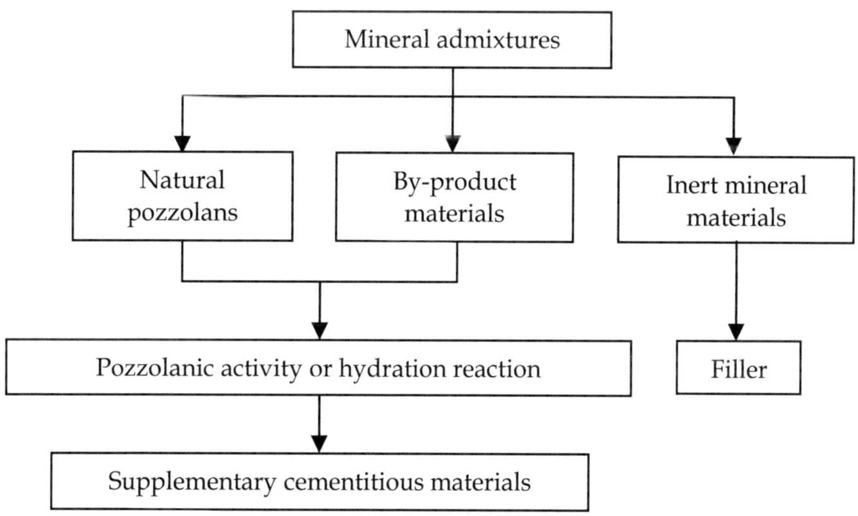

Fig. 2.1. Classification and function of mineral admixtures

Mineral admixtures can be divided into natural pozzolans, by-product materials, and inert mineral materials. Natural pozzolans and by-product materials are generally SCMs [8]. Natural pozzolans are volcanic ashes, diatomaceous earth, calcined clay, metakaolin clay, and rice hull ash. By-product materials include silica fume, fly ash, ggbs, and metakaolin [9-10]. SCMs contribute to the properties of cement-based composites through either pozzolanic activity or hydration reaction [11]. The replacement level varies widely from less than 10 % to more than 50 %, depending on the nature of the SCMs [12]. The physical properties and chemical composition of SCMs are presented in Table 2-1. Silica fume has a large specific surface area and high SiO2 content compared to those of fly ash and ggbs. Hence, the pozzolanic reaction rate of silica fume is considerably higher than that of fly ash or ggbs.

	Portland cement	Fly ash	ggbs	Silica fume
	Physical properties of SCMs			
Specific surface area (m^2/kg)	350-500	300-600	300-500	15000-30000
Bulk density (kg/m^3)	1300-1400	1000	1000-1200	200-300
Specific gravity	3.15	2.30	2.90	2.20
	Chemical compositions of SCMs			
SiO_2	20	50	38	92
Fe_2O_3	3.5	10.4	0.3	1.2
Al_2O_3	5.0	28	11	0.7
CaO	65	3	40	0.2
MgO	0.1	2	7.5	0.2
Na_2O+K_2O	0.5	3.2	1.2	2.0

Table 2.1. Physical properties and chemical composition of SCMs [32]

2.2 Silica fume

Research has indicated that silica fume is suitable for improving the properties of cement-based composites. The advantages of silica fume are illustrated by the following three points [13-15]:

1. Silica fume comprises very fine particles with a specific surface area ranging from 13000 to 30000 m^2/kg. The average particle size is approximately 1/100 that of cement.
2. Silica fume particles are spherical, producing a lubrication effect.
3. Silica fume has a higher silica content, making it a highly effective artificial pozzolan.

Silica fume is a highly reactive pozzolan when used in cement-based composites, reacting with calcium hydroxide to produce additional calcium hydrate. This reaction enhances the mechanical properties and durability of cement-based composites, resulting in stronger, denser, and less permeable materials [16, 17].

The use of silica fume also helps to increase the interfacial bond between paste and aggregate, resulting in an increased tensile strength [16]. Silica fume has a considerable influence on abrasion resistance [18]. Li et al. [19] claimed that replacing cement with silica fume not only reduces the tendency to form cracks, but also decreases the width of cracks in restrained shrinkage test. Due to the increased density of its microstructure, silica fume is highly effective in reducing permeability to water and chloride penetration. The addition of silica fume beyond 5 wt.% of cement significantly reduces electrical conductivity and the addition of silica fume up to 7.5 to 10 wt.% significantly increases corrosion resistance [17].

The combination of silica fume with steel fibers has been shown to enhance compressive strength, splitting tensile strength, abrasion resistance, and impact resistance [20], and aids in the dispersion of fibers in cement-based composites [13,21]. Silica fume even improves the bonding between fibers and mortar [22,23].

2.3 Steel fibers and polyolefin fibers

Fiber reinforced composites are defined as composites incorporating relatively short, discrete, discontinuous fibers. These materials have been added to cement-based composites since 1960. According to ACI 544, there are four categories of fiber reinforced composites: steel fiber, glass fiber, synthetic fiber, and natural fiber reinforced composites.

The use of fibers has been steadily increasing in recent years. Various types of fibers, such as steel, polypropylene, glass, carbon, and cellulose fibers, are used to produce fiber reinforced composites. The properties of various types of fibers are listed in Table 2-2. Steel fibers, in particular, are frequently used in composites. The advantages of adding steel fiber to composites are as follows [24, 25]:

1. Reduced plastic cracking of composites;
2. Resistance to the propagation or growth of cracks;
3. Improved tensile, flexural, and impact strength, as well as increased toughness and energy absorption.

Fibers	Density (g/cm^3)	Tensile strength (MPa)	Young's modulus (GPa)
Carbon fiber	1.81	3.8	228
PVA fiber	1.30	1.6	40
Polypropylene fiber	0.91	0.35-0.50	8.5-12.5
Hooked steel fiber	7.80	1172	200
Glass fiber	0.91-1.03	550-760	--
E-glass fiber	2.53-2.60	3600	75
Polyester fiber	--	875	13
Sisal fiber	1.11-1.37	31-221	15.2

Table 2.2. Properties of various fibers [26-29]

The properties of steel fiber reinforced composites are governed by shape, length, volume fraction, aspect ratio, and surface texture. High volume fraction leads to a decrease in workability and tends to increase material costs [30-32]. Research has suggested that an optimal volume fraction of steel fiber is approximately 2 % [33]. Three types of failure occur in fiber reinforced composites: fiber debonding, fiber pullout, and fiber failure [34]. The use of hooked steel fibers ensures the best possible bonding between the fiber and matrix and maximizing the tensile strength of the steel. The bonding between the fibers and the matrix is strongly influenced by many properties including geometric shape [35].

Polyolefin fibers are a new commercial product manufactured by 3M. Polyolefin fiber enables a high volume of fiber to be used in composites without the occurrence of fiber balling. Polyolefin fibers do not significantly enhance the compressive strength or first-crack flexural strength; however, the presence of polyolefin fibers influences the post crack behavior of composites [36]. In addition, impact resistance and flexural toughness increase as the content of polyolefin fiber increases [37].

A number of studies [36,38,39] have reported that the toughness of polyolefin fiber reinforced concrete is similar to that of steel fiber reinforced concrete. The addition of polyolefin fibers has been shown to increase flexural strength by as much as 13 % and reduce the growth or propagation of cracks by up to 70 % relative to control specimens. In addition, the impact resistance of polyolefin fiber reinforced composites is double that of steel fiber reinforced composites and 14 times greater than composites made without fibers.

2.4 Abrasion resistance

The abrasion of cement-based composites is caused by mechanical contact or exposure to flowing water or particulates. Abrasion results mainly in the localized loss of material from the surface and loosening between the aggregate and paste. The abrasion resistance of cement-based composites is influenced by a number of factors, including compressive strength, properties of the aggregate, water/cementitious ratio, the addition of SCMs, and the properties of supplementary strengthening materials, such as fibers. The ACI committee 201 report indicates that the abrasion resistance of cement-based composites depends primarily upon compressive strength, which is consistent with previous studies [5, 40, 41].

There appears to be a consensus among researchers that compressive strength is a key factor in abrasion resistance; however, opinions vary with regard to the claims of Nanni [42], who suggested that compressive strength is a poor index for evaluating abrasion resistance because it neglects surface and curing conditions. Nanni also reported that the addition of synthetic and steel fibers marginally enhances abrasion resistance. A number of studies have reported that the addition of SCMs, such as fly ash and silica fume, provides a significant increase in abrasion resistance [43-45]. In recent years, fibers and SCMs have been studied individually and in combination to determine their effects on abrasion resistance. The combination of SCMs and fibers has proven particularly interesting, providing considerable scope for further research.

2.5 Concrete abrasion testing

2.5.1 Sand blast abrasion test

Sand blast abrasion testing was conducted in accordance with ASTM C418-05 specifications. This method enables the evaluation of abrasion resistance of cement-based composites subjected to the impingement of air-driven silica sand to determine the abrasion coefficient. The abrasion coefficient, an index of abrasion resistance, is computed as follows:

$$A_c = \frac{V}{A} \tag{2-1}$$

where A is the abraded surface area and V is volume loss through abrasion. A sand blasting cabinet is shown in Fig. 2-2. For the sand blast abrasion test (SBAT) in this study, we prepared ϕ 150 x 64 mm circular discs of each mix.

Fig. 2.2. Sand blasting cabinet

2.5.2 Revolving disk abrasion test

A revolving Disk Abrasion Test (RDAT) was performed in accordance with ASTM C779-05 specifications, procedure A. An RDAT machine, such as the one shown in Fig. 2-3, introduces frictional forces through rubbing and grinding using rotating steel disks in conjunction with abrasive grit. The disks are free floating insofar as they are self supporting, transversely driven along a circular path at 12 rev/min (12 rpm) while individually turned on their own axis at 280 rev/min (280 rpm). In this study, cups attached to the top of the shaft of each disk were loaded with lead shot to produce a uniform total load of 22 N (5 lbf) on each abrading disk face. No. 60 silicon carbide abrasive was fed to the disks at a rate of 4-6 g/min. A test duration of 30 min was generally adequate to produce considerable wear on most concrete surfaces; however, we extended the test period to 60 min to obtain information related to long term abrasion resistance.

By comparing the measurements of average abrasion depth of representative surfaces following 30 and 60 min of exposure, we determined the relative abrasion resistance of these surfaces, using two abrasion coefficients: average abrasion depth and weight loss. In this study, 300 x 300 x 100 (mm) slab specimens were prepared for RDAT.

Fig. 2.3. Revolving disk abrasion test machine

2.5.3 Underwater abrasion test

An underwater abrasion test (UAT) was conducted in accordance with ASTM C1138-05 specifications. The apparatus comprised a drill press, an agitation paddle, a cylindrical steel container housing a disk-shaped concrete specimen, and 65 ±5 steel grinding balls of various sizes. Water in the container was circulated using an immersed agitation paddle powered by the drill press at 1200 rpm. The circulating water, in turn, moved the abrasive charges (steel grinding balls) across the surface of the concrete specimens to produce the abrasion. A total test duration of 24 h was generally adequate to produce significant abrasion in most concrete surfaces. The standard test adopted by this study consists of six 12 h test periods for a total of 72 h. This study prepared ϕ 300 x 100 mm circular discs for UAT.

The volume of the specimens is calculated as follows:

$$V_t = \frac{W_{air} - W_{water}}{G_w} \tag{2-2}$$

where V_t is the volume of the specimen, W_{air} is the weight of the specimen in air, W_{water} is the weight of the specimen in water and G_w is the unit weight of water.

The volume of concrete lost at the end of any time increment of testing is calculated as follows:

$$VL_t = V_i - V_t \tag{2-3}$$

where VL_t is the volume of material lost through abrasion and Vi is the volume of the specimen prior to testing.

In addition, the average depth of abrasion at the end of any time increment of testing is calculated as follows:

$$ADA_t = \frac{VL_t}{A} \tag{2-4}$$

where ADA_t is the average depth of abrasion and A is the surface area of the top of the specimens. In this test, the volume of material lost and the average abrasion depth are used as indices of abrasion resistance.

2.5.4 Los Angeles abrasion test

The Los Angeles abrasion test (LAAT) was conducted in accordance with of ASTM C131-06 specifications, of which LAAT is a modified version. The machine comprises a steel cylinder, into which ϕ100 x 200 mm specimens are placed with eight steel spheres. Following the rotation of the samples, the percentage loss (difference between the original mass and the final mass of the specimen) was calculated and used as an index of abrasion resistance.

3. Results and discussion

This study used two types of fiber (steel fiber and polyolefin fiber), three silica fume contents (0 %, 5 % and 10 % by weight of cement), four steel fiber dosages (0 %, 0.5 %, 1.0 % and 2.0 % by volume of cement-based composites), and two water/cementitious ratios (0.35 and 0.55) in the mix designs. We also investigated rock wool waste and the shape of fibers and particles, as they pertain to abrasion resistance.

3.1 Sand blasting abrasion test

The abrasion coefficient of specimens decreased with an increase in fiber and silica fume content, as illustrated in Fig. 3-1. The inclusion of 0.5 vol. % steel fibers in the composites decreased the abrasion coefficient considerably. However, the abrasion coefficient did not significantly change when the steel fiber content exceeded 1.0 vol. %. The abrasion coefficient also decreased with an increase in silica fume content for the specimens for both w/cm ratios. The addition of steel fibers produced a denser, stronger surface, resulting in superior resistance to wear. The abrasion coefficient of specimens containing 10 % silica fume and 2.0 % steel fibers with w/cm ratios of 0.35 and 0.55 were 2.21×10^{-3} and 3.21×10^{-3} (cm^3/cm^2), respectively. These results are approximately 50 % lower than those of the control specimen. Composites with 10 % silica fume and 2.0 % fiber also exhibited excellent abrasion resistance.

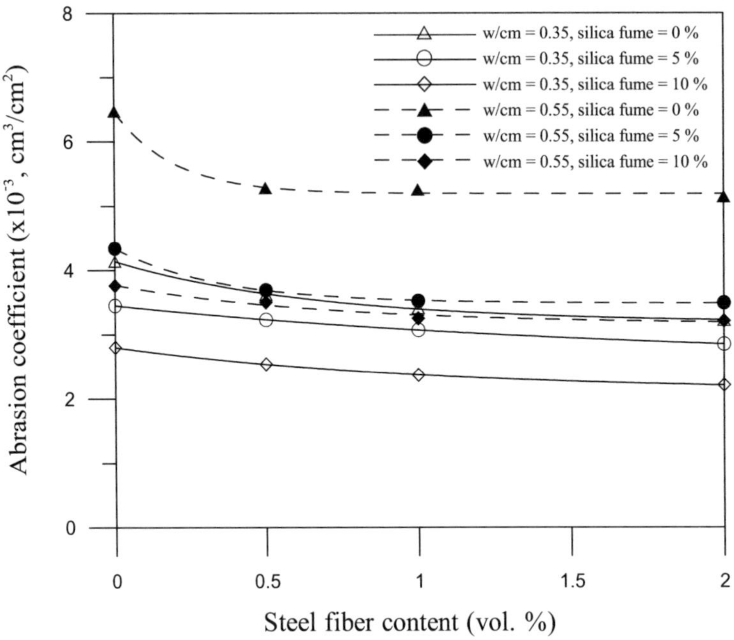

Fig. 3.1. Abrasion coefficient vs. steel fiber content curves

The abrasion coefficient appears to decrease exponentially with compressive strength as illustrated in Fig. 3-2. Clearly, compressive strength has more of an influence on the abrasion coefficient of specimens with higher w/cm than on those with lower w/cm.

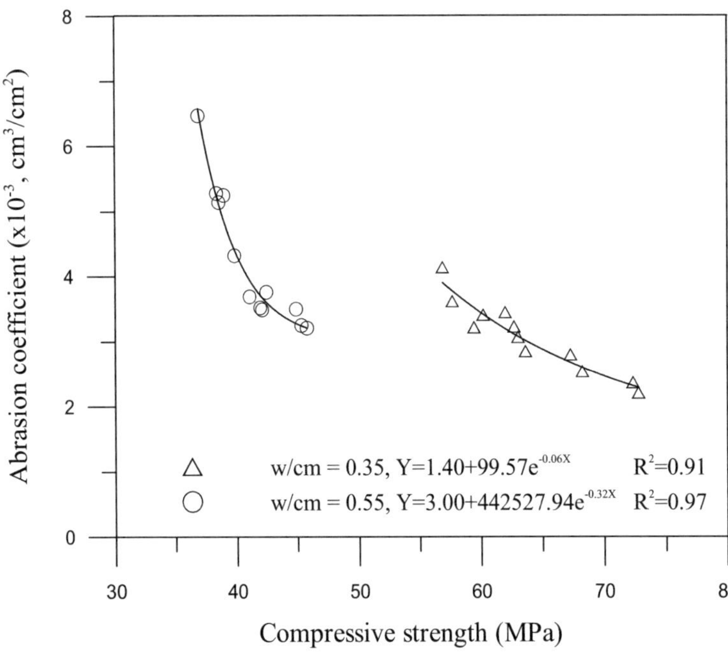

Fig. 3.2. Abrasion coefficient vs. compressive strength curves

The mechanical properties of steel fiber reinforced composites with silica fume are influenced by many factors. Multiple regression analysis was conducted to evaluate the influence and interaction of the material variables, through the selection of the following independent variables: water/cementitious ratio (*w/cm*), silica fume content (W_s) and steel fiber content (V_f) and the dependent variable, abrasion coefficient (A_c). To determine the effects of the interactions among all factors related to these materials, we assumed the following regression model:

$$\begin{aligned} Y &= a1 + a2(w/cm) + a3(W_s) + a4(V_f) + a5(w/cm)(W_s) \\ &+ a6(w/cm)(V_f) + a7(W_s)(V_f) + a8(w/cm)(W_s)(V_f) \end{aligned} \tag{3-1}$$

where Y is the predicted value and w/cm, W_s, V_f are independent variables, a1 is the intercept and a2, a3, a4, a5, a6, a7, and a8 are the regression coefficients. The results of multiple regression analysis are presented below.

$$\begin{aligned} A_c &= 0.88 + 8.88(w/cm) - 0.07(W_s) - 0.17(V_f) \\ &-0.56(w/cm)(W_s) - 0.66(w/cm)(V_f) - 0.10(W_s)(V_f) \\ &+0.70(w/cm)(W_s)(V_f) \end{aligned} \tag{3-2}$$

As illustrated in Fig. 3-3, the experimental abrasion resistance was found to be strongly correlated with the estimated values, with a correlation coefficient of 0.91. The estimated values are plotted against the experimental values in Figs. 3-3. The deviation is defined as the distance of each point on the plot from the diagonal line. In general, points are uniformly scattered around the diagonal line for all assumed models. The above analysis indicates that the model parameters are strongly dependant on the mechanical properties tested, providing valuable information with which to predict the mechanical properties of steel fiber reinforced composites.

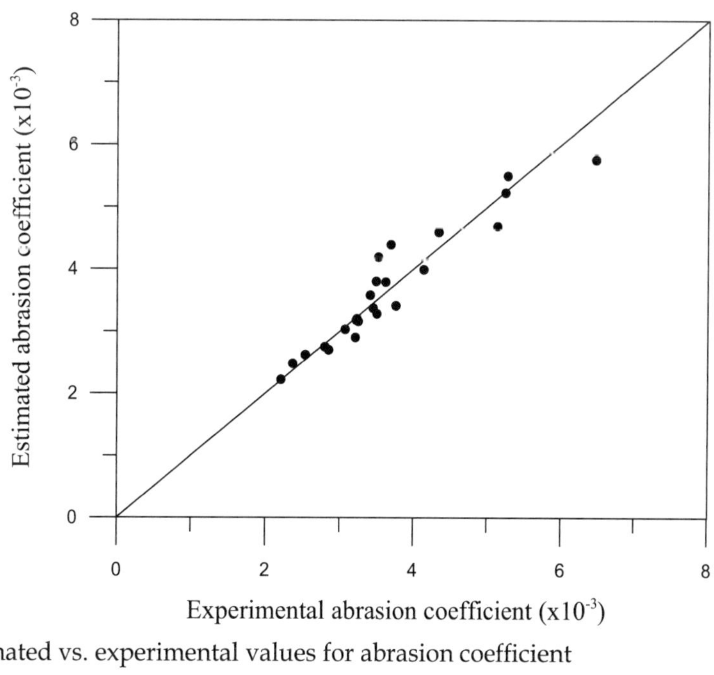

Fig. 3.3. Estimated vs. experimental values for abrasion coefficient

(a) x30

(b) x500

(c) x20000

Fig. 3.4. SEM images of rock wool waste

Rock wool waste is an inorganic fibrous substance produced by steam blasting and cooling molten glass. Rock wool waste obtained from thermal insulation materials is crushed and ground. Like other by-product materials, rock wool waste can be used as coarse aggregates, fine aggregates, cementitious material, or inert fillers in concrete, depending on its chemical composition and particle size. Figure 3-4 presents an SEM image of rock wool waste displaying cylindrical and fiber shapes.

The abrasion coefficient of specimens decreased with an increase in rock wool content (Fig. 3-5). The abrasion coefficients of specimens containing 10 wt. % rock wool waste at w/cm ratios of 0.55 and 0.65 were as much as 4 % and 5 % lower than control specimens, respectively. This indicates that when rock wool is used, abrasion resistance is strongly associated with the bond between the cement paste and fine aggregate. Finer rock wools enhances abrasion resistance, providing considerable performance benefits for cement-based composites.

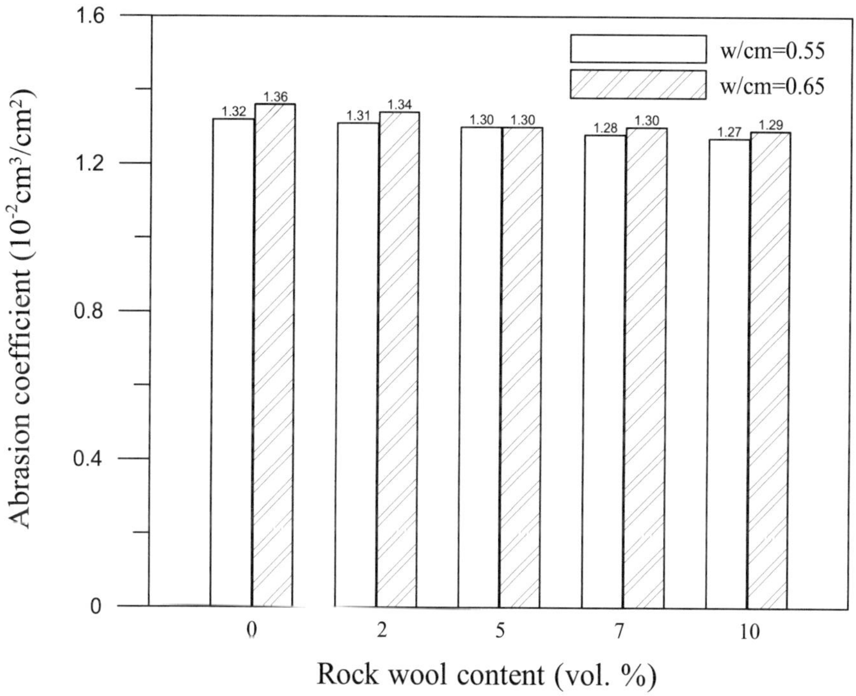

Fig. 3.5. Abrasion coefficient vs. rock wool content histograms

3.2 Los Angeles abrasion test

Composite samples aged 28 days were tested using the Los Angeles abrasion test, and the weight loss following 100, 300, and 500 rotations is summarized in Table 3-1. The relationships between weight loss and rotations for samples with w/cm ratios of 0.35 and 0.55 are shown in Figs. 3-6 and 3-7. As seen in the figures, weight loss and abrasion resistance index significant increased with an increase in the number of rotations and the w/cm ratio. The BF10 specimen demonstrated the greatest weight loss, due to balling and the poor dispersion of fibers, which may have been the result of weak bonding between the fibers and paste in partial cement-based composites, particularly at higher w/cm ratios. These results indicate that the scaling of composites may increase weight loss. Conversely, silica fume helps to disperse fibers uniformly throughout composites, thereby significantly improving abrasion resistance. Silica fume with a particle size of approximately 0.2 μm appears to improve the dispersal of fibers throughout the matrix. According to the testing results, specimens containing silica fumes demonstrated lower weight loss than control specimens. Specimens combining silica fume and polyolefin fibers demonstrated performance superior to that of specimens containing either silica fume or fibers. In conclusion, an appropriate combination of silica fume and fibers provides the highest abrasion resistance, from which we infer that silica fume not only helps to disperse fibers but also strengthens the bond between fibers and the cement-based matrix.

	Weight loss (%)		
Mix no.	100 rotations	300 rotations	500 rotations
A	4.22%	8.18%	11.34%
B	3.21%	14.88%	21.00%
AS5	3.35%	7.73%	11.25%
BS5	4.43%	11.15%	17.06%
APS	3.42%	7.65%	10.03%
BPS	4.37%	12.36%	17.92%
AS5PS	2.07%	5.93%	10.54%
BS5PS	1.97%	6.37%	14.25%
AF05	2.14%	4.51%	9.91%
BF05	2.11%	21.89%	27.42%
AF10	1.49%	3.37%	12.08%
BF10	2.40%	11.88%	18.59%

Note: A represents w/cm = 0.35
B represents w/cm = 0.55
S5 represent silica fume = 5 wt. %
PS represent polyolefin fiber
F05 represent steel fiber = 0.5 vol. %
F10 represent steel fiber = 1.0 vol. %

Table 3.1. Weight loss following Los Angeles abrasion testing

Weight loss of specimens following 100, 300, and 500 rotations is shown in Figs. 3.8-3.10, respectively. The weight loss of specimens decreased in proportion to the addition of polyolefin fibers, steel fibers, or silica fume. This tendency is more obvious at 100 rotations than at 300 or 500 rotations, due to the poor bonding caused by composite scaling. Poor bonding could be overcome by long-term curing and the addition of SCMs. Control specimens also displayed severe weight loss, which may be due to the fact that the control specimens had much lower compressive strength, consistent with the results of SBAT. The combination of silica fume and steel fibers or polyolefin fibers is more effective in enhancing abrasion resistance, indicating that this combination enhances bond strength and compressive strength. In summary, bond strength between fibers and pastes is another important factor (in addition to compressive strength) in abrasion testing.

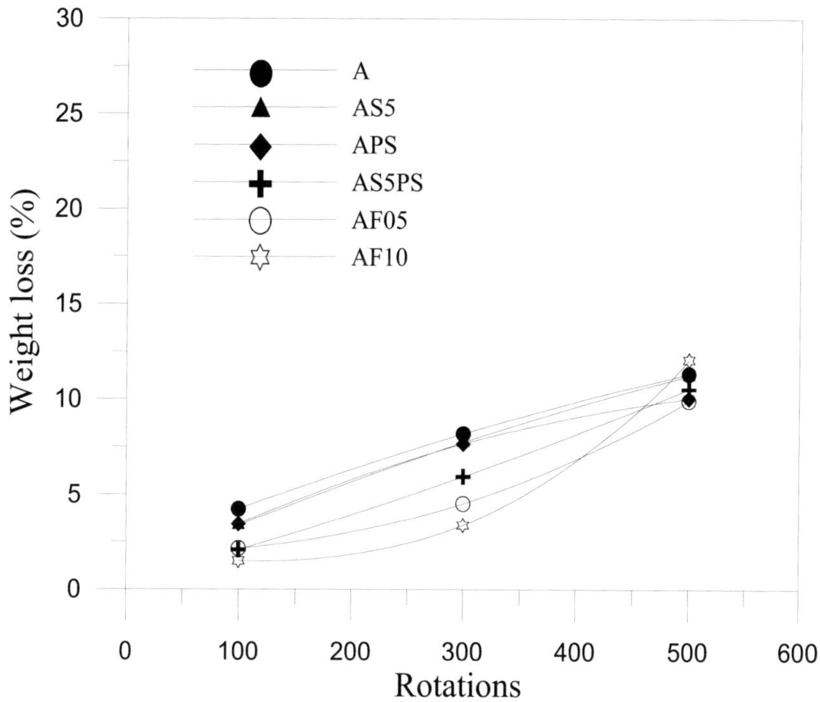

Fig. 3.6. Weight loss vs. rotation curves (w/cm = 0.35)

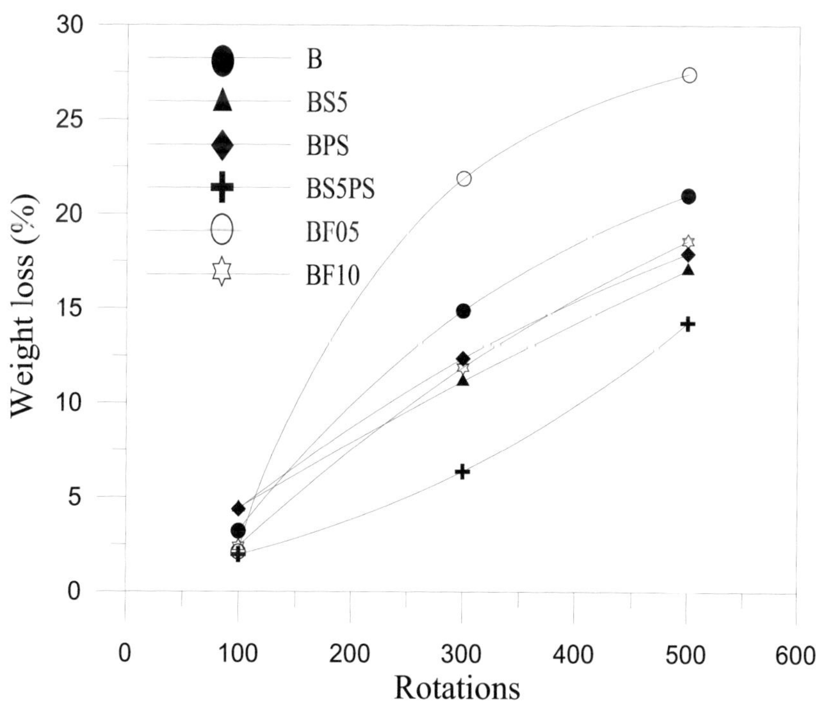

Fig. 3.7. Weight loss vs. rotation curves (w/cm = 0.55)

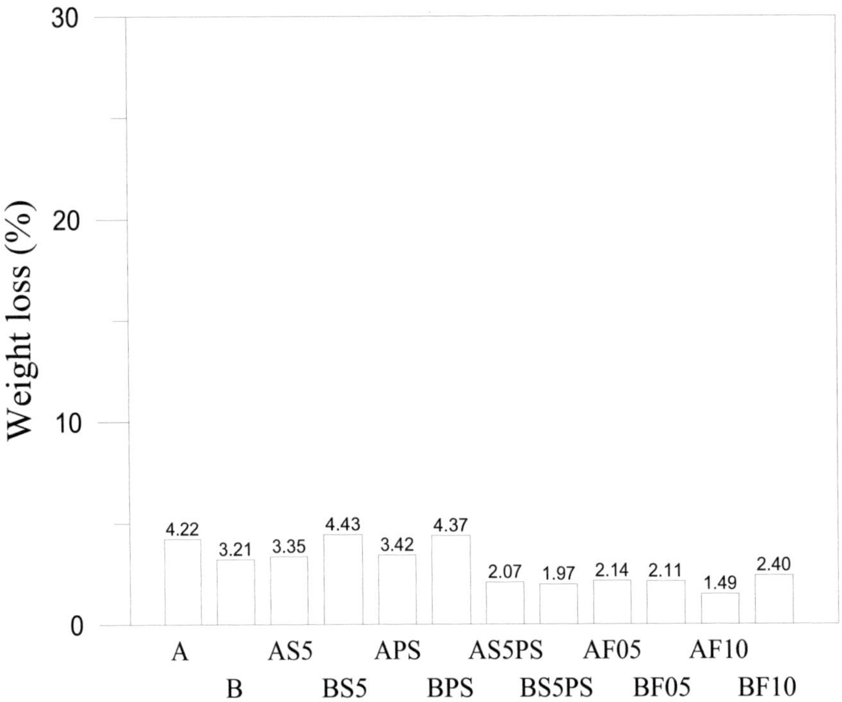

Fig. 3.8. Weight loss histogram following 100 rotations

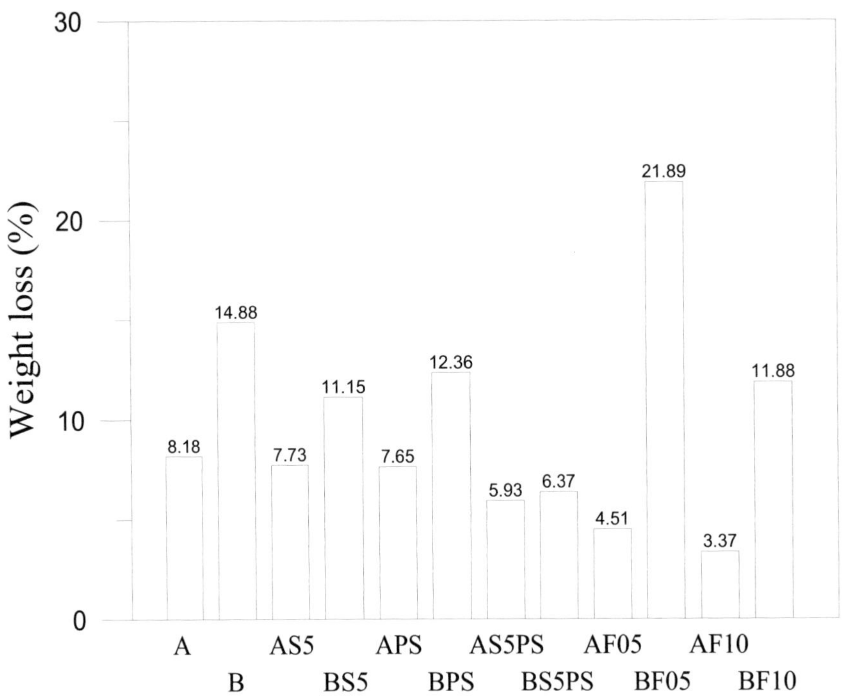

Fig. 3.9. Weight loss histogram following 300 rotations

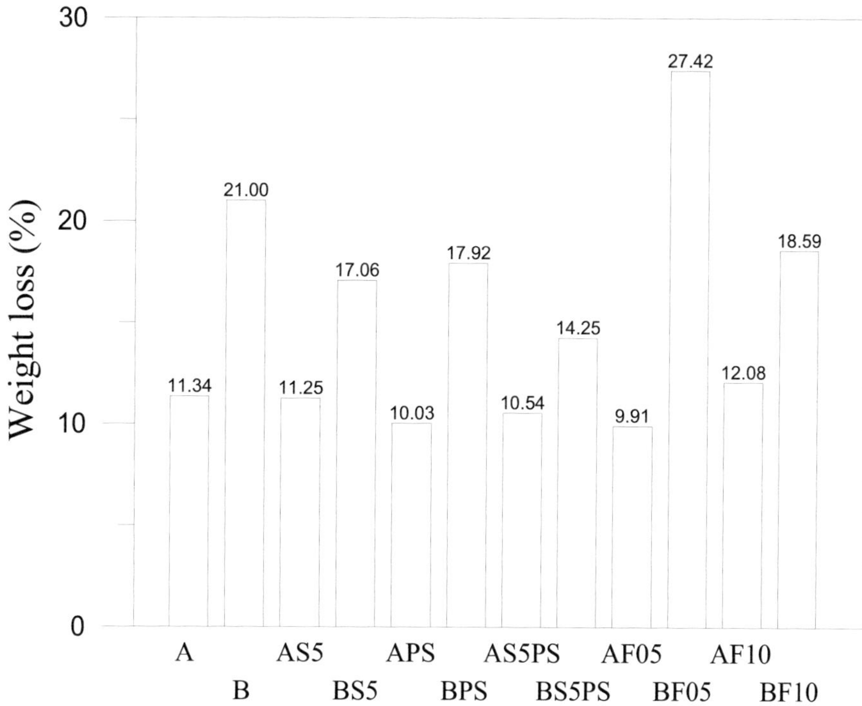

Fig. 3.10. Weight loss histogram following 500 rotations

4. Conclusions

According to the previous results and discussion, our conclusions are presented as follows.

1. Abrasion resistance is clearly influenced by the addition of silica fume and steel fibers. The inclusion of silica fume in composites increases abrasion resistance (-32~-42 %), by increasing the dense hydrated calcium silicate to provide a more refined pore system. The inclusion of steel fibers marginally influences abrasion resistance. The combination of steel fibers and silica fume provides little improvement in abrasion resistance (-8~-15 %).
2. Based on the correlation coefficient of statistic analysis, the abrasion resistance of cement-based composites is strongly correlated with w/cm and silica fume and steel fiber content.
3. The addition of rock wool waste enhances abrasion resistance in a manner similar to that of SCMs in cement-based composites.
4. The modified Los Angeles abrasion test used in this study is suitable for the evaluation of abrasion resistance in cement-based composites.
5. The inclusion of silica fume in cement-based composites results in a denser microstructure with fewer pores, thereby enhancing abrasion resistance. The inclusion of silica fume also enhances the bond between fibers and paste. Specimens combining silica fume with polyolefin fibers demonstrated superior abrasion resistance compared to specimens containing individual constituents of silica fume or fibers.
6. Bond strength between fibers and paste is another important factor (in addition to compressive strength), associated with abrasion resistance.

5. References

[1] B.H. Oh, S.W. Cha, B.S. Jang, S.Y. Jang, Development of high-performance concrete having high resistance to chloride penetration, Nuclear Engineering and Design (Switzerland) , Vol. 212 No. 1–3, 2002, pp. 221-231.

[2] M. Sahmaran, V.C. Li, Influence of microcracking on water absorption and sorptivity of ECC, Materials and structures, Vol. 42, No. 5, 2009, pp. 593-603.

[3] H.T. Antoni, N. Saeki, Performance of FRC against chloride penetration under loading. Proceedings of the JCI, Vol. 26, No. 1, 2004, pp. 921-926.

[4] P.K. Chang, Y.N. Peng, C.L. Hwang, A design consideration for durability of high-performance concrete. Cement and Concrete Composites, Vol. 23, No. 4-5, 2001, pp. 375-380.

[5] P.K. Mehta, Concrete: structure, properties, and materials. Englewood Cliffs, New Jersey: Prentice-Hall, 1993.

[6] J. Rapoport, C.M. Aldea, S.P. Shah, B. Ankenman, A. Karr, Permeability of cracked steel fiber-reinforced concrete. Journal of Materials in Civil Engineering, Vol. 14, No. 4, 2002, pp. 355-358.

[7] American Concrete Institute, Guide for Conducting a Visual Inspection of Concrete in Service, ACI-201.1R-08, ACI Committee Report, 2008.

[8] V.G. Papadakis, S. Tsimas, Greek supplementary cementing materials and their incorporation in concrete. Cement and Concrete Composites, Vol. 27, No. 2, 2005, pp. 223-231.

[9] O.S.B. Al-Amoudi, A.A. Almusallam, M.M. Khan, M. Maslehuddin, Effect of hot weather on compressive strength of plain and blended cement mortars. Proceedings of the 4th Saudi engineering conference, King Abdulaziz University Jeddah, vol. 2, 1995, pp. 193-199.

[10] O.S.B. Al-Amoudi, M. Maslehuddin, S.N. Abduljauwad, Influence of sulfate ions on chloride-induced reinforcement corrosion in plain and blended cement concretes. Cement and Concrete Aggregates, Vol. 16, No. 1, 1994, pp. 3-11.

[11] P. Lawrence, M. Cyr, E. Ringot, Mineral admixtures in mortars effect of type, amount and fineness of fine constituents on compressive strength. Cement and Concrete Research, Vol. 35, No. 6, 2005, pp. 1092-1105.

[12] J. Newman, B.S. Choo, Advanced Concrete Technology: Constituent Materials. Butterworth-Heinemann Ltd., 2003.

[13] D.D.L. Chung, Review-Improving cement-based materials by using Silica Fume. Journal of Materials Science, Vol. 37, No. 4, 2002, pp. 673-682.

[14] O.S.B. Al-Amoudi, M. Maslehuddin, M.A. Bader, Characteristics of silica fume and its impact on concrete in the Arabian Gulf. Concrete and Construction, Vol. 35, No. 2, 2001, pp. 45-50.

[15] O.S.B. Al-Amoudi, M. Maslehuddin, M. Shameem, M. Ibrahin, Shrinkage of plain and silica fume cement concrete under hot weather. Cement and Concrete Composites, Vol. 29, No. 9, 2007, pp. 690-699.

[16] ACI Committee 234, Guide for the use of silica fume in concrete (ACI 234-06). American Concrete Institute, Farmington Hills, 2006.

[17] D. Sideny, S. Sadananda, Densified silica fume: particle sizes and dispersion in concrete. Materials and Structures, Vol. 39, No. 293, 2006, pp. 849-859.

[18] P.C. Laplante, P.C. Aitcin, D. Vezina, Abrasion resistance of concrete. Journal of Materials in Civil Engineering, Vol. 3, No. 1, 1991, pp. 19-28.
[19] Z. Li, M. Qi, B. Ma, Crack width of high-performance concrete due to restrained shrinkage. Journal of Materials in Civil Engineering, Vol. 11, No. 3, 1999, pp. 214-223.
[20] O. Eren, K. Marar, T. Celik, Effects of silica fume and steel fibers on some mechanical properties of high-strength fiber-reinforced concrete. Journal of Testing and Evaluation, Vol. 27, No.6 , 1999, pp. 380-387.
[21] P.W. Chen, X. Fu, D.D.L. Chung, Microstructural and mechanical effects of latex, methyleellulose and silica fume on carbon fiber reinforced cement. ACI Materials Journal, Vol. 94, No. 2, 1997, pp. 147-155.
[22] X. Fu, D.D.L. Chung, Effects of water-cement ratio, curing age, silica fume, polymer admixtures, steel surface treatments, and corrosion on bond between concrete and steel reinforcing bars. ACI Materials Journal, Vol. 95, No. 6, 1998, pp. 725-734.
[23] H. Yan, W. Sun, H. Chen, Effect of silica fume and steel fiber on the dynamic mechanical performance of high-strength concrete. Cement and Concrete Research, Vol. 29, No. 3, 1999, pp. 423-426.
[24] J. Newman, B.S. Choo, Advanced Concrete Technology: Processes. Butterworth-Heinemann Ltd., 2003.
[25] J.I. Daniel, J.J. Roller, E.D. Anderson, Fiber reinforced Concrete, Portland Cement Association, 1998, pp. 22-26.
[26] J. Kaufmann, D. Hesselbarth, High performance composites in spun-cast elements. Cement and Concrete Composites, Vol. 29, No. 10, 2007, pp. 713-722.
[27] M.J. Shannag, S.A. Al-Ateek, Flexural behavior of strengthened concrete beams with corroding reinforcement. Construction and Building Materials, Vol. 20, No. 9, 2006, pp. 834-840
[28] J. Péra, J. Ambroise, Fiber-reinforced Magnesia-phosphate Cement Composites for Rapid Repair. Cement and Concrete Composites, Vol. 20, No. 1, 1998, pp. 31-39.
[29] G. Ramakrishna, T. Sundararajan, Studies on the durability of natural fibres and the effect of corroded fibres on the strength of mortar. Cement and Concrete Composites, Vol. 27, No. 5, 2005, pp. 575-582.
[30] M. Maalej, T. Hashida, V. Li. Effect of fiber volume fraction on the off-crack-plane fracture energy in strain hardening engineered cementitious composites. Journal of the American Ceramic Society, Vol. 78, No. 12, 1995, pp. 3369-3375.
[31] N. Banthia, M. Sappakittipakorn, Toughness enhancement in steel fiber reinforced concrete through fiber hybridization. Cement and Concrete Research, Vol. 37 No. 9, 2007, pp. 1366-1372.
[32] P.W. Chen, D.D.L. Chung, Low-drying-shrinkage concrete containing carbon fibers. Composites Part B: Engineering, Vol. 27, No. 3-4, 1996, pp. 269-274.
[33] A.E. Naaman, Engineered steel fibers with optimal properties for reinforcement of cement composites. Journal of Advanced Concrete Technology, Vol. 1, No. 3, 2003, pp. 241-252.
[34] R.F. Zollo, Fiber-reinforced Concrete: an Overview after 30 Years of Development. Cement Concrete Composites, Vol. 19, No. 2, 1997, pp. 107-122.
[35] V.C. Li, H. Stang, Interface property characterization and strengthening mechanisms in fiber reinforced cement based composites. Advanced Cement Based Materials, Vol. 6, No. 1, 1997, pp. 1-20.

[36] B.D. Neeley, E.F. O'Neil, Polyolefin fiber reinforced concrete. Proceedings of the Materials Engineering Conference, Vol. 1, Materials for the New Millennium, 1996, pp. 113-122.

[37] A. Tagnit-Hamou, Y. Vanhove, N. Petrov, Microstructural analysis of the bond mechanism between polyolefin fibers and cement pastes. Cement and Concrete Research, Vol. 35, No. 2, 2005, pp. 364-370.

[38] http://www.3m.com/

[39] V. Ramakrishnan, Performance characteristics of polyolefin fiber reinforced concrete. Proceedings of the Materials Engineering Conference, Vol. 1, Materials for the New Millennium, 1996, pp. 93-102.

[40] P. Laplante, P.C. Aifcin, D. Vezina, Abrasion Resistance of Concrete. Journal of Materials in Civil Engineering, Vol. 3, No. 1, 1991, pp. 19-28.

[41] K. Andrej, M. Matjaz, S. Jakob, P. Igor, Abrasion Resistance of Concrete in Hydraulic Structures. ACI materials journal, Vol. 106, No. 4, 2009, pp. 349-356.

[42] A. Nanni, Abrasion Resistance of Roller Compacted Concrete. ACI Materials Journal, Vol. 86, No. 53, 1989, pp. 559-565.

[43] B.W. Langan, R.C. Joshi, M.A. Ward, Strength and Durability of Concrete Containing 50% Portland Cement Replacement by Fly Ash and other Materials. Canadian Journal of Civil Engineering, Vol. 17, 1990, pp. 19-27.

[44] P.J. Tikalsky, P.M. Carrasquillo, R.L. Carrasquillo, Strength and Durability Considerations Affecting Mix Proportioning of Concrete Containing Fly Ash. ACI Materials Journal, Vol. 85, No. 6, 1988, pp. 505-511.

[45] W.T. Lin, R. Huang, C.L. Lee, H.M. Hsu, Effect of Steel Fiber on the Mechanical Properties of Cement-based Composites Containing Silica Fume. Journal of Marine Science and Technology, Vol. 16, No. 3, 2008, pp. 214-221.